Chemical Sciences in Early Drug Discovery

Discovery

Medicinal Chemistry 2.0

Chemical Sciences in Early Drug Discovery

Medicinal Chemistry 2.0

Pierfausto Seneci
Department of Chemistry
University of Milan
Milan, Italy

ELSEVIER

Elsevier
Radarweg 29, PO Box 211, 1000 AE Amsterdam, Netherlands
The Boulevard, Langford Lane, Kidlington, Oxford OX5 1GB, United Kingdom
50 Hampshire Street, 5th Floor, Cambridge, MA 02139, United States

Notices
Knowledge and best practice in this field are constantly changing. As new research and experience broaden our understanding, changes in research methods, professional practices, or medical treatment may become necessary.

Practitioners and researchers must always rely on their own experience and knowledge in evaluating and using any information, methods, compounds, or experiments described herein. In using such information or methods they should be mindful of their own safety and the safety of others, including parties for whom they have a professional responsibility.

To the fullest extent of the law, neither the Publisher nor the authors, contributors, or editors, assume any liability for any injury and/or damage to persons or property as a matter of products liability, negligence or otherwise, or from any use or operation of any methods, products, instructions, or ideas contained in the material herein.

Library of Congress Cataloging-in-Publication Data
A catalog record for this book is available from the Library of Congress

British Library Cataloguing-in-Publication Data
A catalogue record for this book is available from the British Library

ISBN: 978-0-08-099420-8

For information on all Elsevier publications
visit our website at https://www.elsevier.com/books-and-journals

Working together
to grow libraries in
developing countries

www.elsevier.com • www.bookaid.org

Publisher: Susan Dennis
Acquisition Editor: Emily McCloskey
Editorial Project Manager: Billie Jean Fernandez
Production Project Manager: Paul Prasad Chandramohan
Cover Designer: Mark Rogers

Typeset by SPi Global, India

Contents

Introduction   vii

1.   Step I: Target Identification

1.1   The Foundation: Molecular Biology, No Chemistry   2
1.2   Chemistry in Modern Target Identification: Phenotypic
      Screening, Forward Chemical Genetics   9
1.3   Kinesin Spindle Protein (KSP, Eg5)—Monastrol   12
      References   23

2.   Step II: Target Validation

2.1   The Foundation: Molecular Biology, Once More
      No Chemistry   34
2.2   Chemistry in Modern Target Validation: Phenotypic
      Screening, Reverse Chemical Genetics   35
2.3   Bromodomain and Extraterminal Domain (BET)
      Readers—I-BET (GSK525762A)   56
      References   64

3.   Step IIIa: Biological Hit Discovery Through High-Throughput Screening (HTS): Random Approaches and Rational Design

3.1   The Foundation: Biology-Oriented High-Throughput
      Screening (HTS), No Chemistry   79
3.2   Virtual High-Throughput Screening (vHTS)   83
3.3   Biophysical Screening in Fragment-Based Drug Discovery   98
      References   104

4. Step IIIb: The Drug-Like Chemical Diversity Pool: Diverse and Targeted Compound Collections

4.1	The Foundation: Drug-Like Chemical Diversity	117
4.2	Heat Shock Protein 90 (Hsp90): Hit Discovery by HTS, vHTS, and FBDD	124
4.3	Expanding the Drug-Like Space out of the RO5 Box	135
	References	162
	Further Reading	177

Index	179

Introduction

Why another book series dealing with Medicinal Chemistry? High-quality, recently published volumes written by experts from the public/academic and the private/industrial sector already exist.

Medchem books belong to two main categories. Multiauthored volumes enjoy excellence in each Chapter and Section, as contributions are "confined" to the main area of expertise of each author. Such books are mostly aimed toward seasoned, chemistry-oriented researchers in pharmaceutical R&D, although they contain student-friendly introductions on the how-what-when of medchem approaches in drug discovery. Their contents are sometimes overlapping, due to multiple authorship, while their quality depends on the reputation of the main Editor. I could obviously not compare with the likes of Wermuth[1] and Williams[2] in recruiting authors for specific Chapters: such a book structure was not a valid option for me—anyway, I had a different one in mind.

A "One-man show" presents the audience (Ph.D. students primarily, but also medchem/drug discovery scientists) with a coherent view on medicinal chemistry, and on its features. Such a volume is suited to be a textbook, and must cover either the basics of medchem, and its connections to the whole pharmaceutical R&D. Wonderful medchem textbooks are available: my picks would include Silverman[3] and Patrick.[4]

If both options—in particular the latter—are adequately covered, why should I add to it? This book, and its future sequels to cover the whole pharma R&D process, should provide the Reader with an original viewpoint, and—rather than substituting the mentioned textbooks—should integrate them, and be useful to students and experienced medicinal chemists.

It is my belief that medicinal chemistry and medicinal chemists should contribute to any phase/process belonging to pharmaceutical R&D. Traditional medchem treatises do not extensively cover early steps in drug discovery, as they are considered the biologists' playground. These subjects are the focus of this book, because chemical compounds (i.e., medicinal chemists) heavily influence their outcome in modern drug discovery.

Medchem permeates pharmaceutical R&D from A (identifying a molecular target, step 1) to Z (supplying industrial quantities of a drug to the market, step 9, Fig. 1). Lab R&D projects do not usually require all these steps: me-too drugs are targeted against validated, MoA-characterized targets, and rational design and HTS are seldom pursued in the same project as a source of hits.

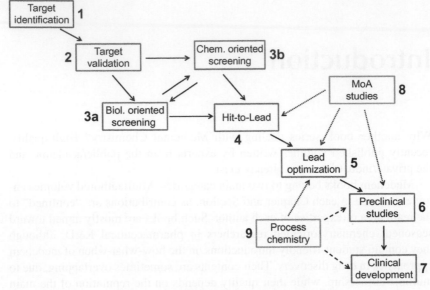

FIG. 1 The R&D pharmaceutical process.

Nevertheless, a modern medicinal chemist should be acquainted with all of them, to extract therapeutic potential from a molecular target and/or a compound class.

My goal is a "matrix" coverage of medchem-related drug R&D. In this first volume, four key steps in the pharmaceutical R&D process are covered in four Chapters (Fig. 1). *Target identification-TI*-step 1 describes the identification of disease-associated molecular targets (usually proteins). *Target validation-TV*-step 2 connects at an early stage the chosen target and the pursued disease(s). *Hit discovery-HD-step 3* is split into assay development/high-throughput screening (HTS, step 3a) and compound collections (step 3b, Fig. 1). The former describes suitable assays to identify target-directed or phenotype-selected chemical diversity (CD); the latter covers the rational selection and assembly of HTS collections—either made by drug-like/RO5-compliant or by natural product-like chemical diversity.

Each Chapter deals with the foundation of each step (modern-classical approaches, expected results); with traditional medchem contributions (if any) for each step; and with medchem-driven innovation (conceptual and technological improvements, leading to quantifiable results). Such examples illustrate how a medicinal chemist can contribute to the first steps of the R&D process.

I believe that there's enough here to grant a textbook label for this series. Medchem Ph.D. students should find here the foundation of biologically active compounds and of medicinal chemistry, and the description of each pharmaceutical R&D phase. The innovation content should make the students aware of trends in medchem-oriented pharmaceutical R&D, and should be useful for seasoned medicinal chemists with an open mind toward the future and its opportunities. We, as modern medicinal chemists, are key to enable the discovery and development of novel drugs. Let's have our fellow biologists-pharmacologists-clinicians realize that, by (also) supporting their work in areas that (apparently, but erroneously) are believed to be "medchem-free"!

REFERENCES

1. Wermuth, C.; Aldous, D.; Raboisson, P.; Rognan, D., Eds.; The Practice of Medicinal Chemistry; 4th ed. Academic Press: Oxford, 2015, 902 p.
2. Lempke, T. L.; Williams, D. A.; Roche, W. F.; Zito, S. W., Eds.; Foye's Principles of Medicinal Chemistry; 7th ed. Lippincott Williams & Wilkins: Philadelphia, 2013, 1500 p.
3. Silverman, R. B.; Holladay, M. W. The Organic Chemistry of Drug Design and Drug Action; 3rd ed.; Academic Press: San Diego, 2014, 517 p.
4. Patrick, G. L. An Introduction to Medicinal Chemistry; 5h ed.; Oxford University Press: Oxford, 2013, 789 p.

Chapter 1

Step I: Target Identification

Molecular targets and their identification are the foundation of R&D pharmaceutical projects (step 1, Fig. 1.1). No matter what is done later, choosing a wrong target dooms any R&D project. Conversely, a disease-related target leads—if properly characterized for its in vitro and in vivo physiological and pathological role, and modulated with drug-like biological or chemical entities—to therapeutically relevant outcomes.

The definition of "druggable genome" dates back to 2002.[1] If at least one gene family member interacts with one or more biological or chemical compounds, the whole gene family is druggable. A ≈3000 druggable gene number was predicted.[1] A similar size was estimated by large-scale gene knockout studies for disease-related, putative drug target genes,[2] i.e., ≈10% of the estimated 30K gene-sized human genome possessing putative disease-modifying features.[3] The overlapping population of druggable, disease-related genes was then set between 600 and 1500 (Fig. 1.2, top).

Today's view on druggable genome and drug targets has changed. The size of the human genome is smaller (≈20–25K genes),[4] but the recent modulation/"druggability" of therapeutically relevant protein-protein interactions[5] and the success of biological drugs[6] has increased the druggable genome size to ≈5K genes. Reliable and efficient validation tools (see next chapter) have increased the number of disease-related molecular targets—once more, to ≈5K targets. Thus, a larger set of overlapping druggable, disease-related targets exists—let's say ≥2K targets (Fig. 1.2, bottom). This size should further increase in future due to druggability- and target validation-related innovations.

An approved drug gives full confidence in its molecular target. A rigorous analysis[7] identified 555 gene targets of past and present drugs. Drug candidates undergoing clinical trials inspire similar confidence for their targets. The same report[7] listed 475 gene targets of clinically tested compounds (small molecules or biologicals).

Large pharmaceutical companies ("big pharmas") usually work on "me too" candidates[8] acting on market- or clinically validated targets. Their risk of failure is lower, while their market potential—powered by the marketing capacity of big pharmas—remains significant.

Even if each of the 475 targets in clinical validation would see the approval of one of its modulators as drugs, the 1030 validated targets would represent

Chemical Sciences in Early Drug Discovery. https://doi.org/10.1016/B978-0-08-099420-8.00001-8

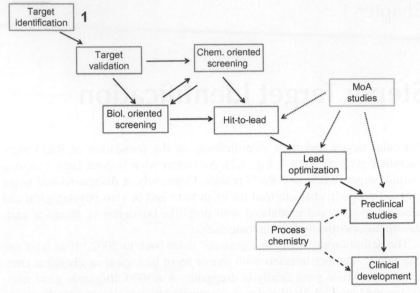

FIG. 1.1 The R&D pharmaceutical process.

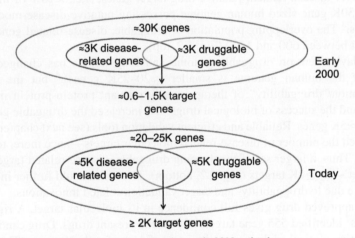

FIG. 1.2 The druggable genome: size estimation, early 2000 and today.

≈50% of druggable, disease-related genes. The identification and validation of the remaining targets will benefit from chemistry-based approaches and tools.

1.1. THE FOUNDATION: MOLECULAR BIOLOGY, NO CHEMISTRY

Gene-encoded protein targets stem from the unraveling of DNA structure in 1953,[9] of three RNA codons in 1962,[10] and from DNA sequencing in 1977.[11] Traditional medicine remedies,[12] such as aspirin,[13] were made available to patients much earlier. The molecular targets of most *pre*-Watson and Crick drugs were identified after their therapeutic application. Even today the lack of a mechanism of action (MoA) does not prevent the development of a drug, if potent and safe in humans.[14] A known MoA, though, explains preclinical and clinical effects of drug candidates, and supports their postmarketing surveillance among patients.

A molecular target-driven working hypothesis is important to any pharma R&D project, even at the earliest stage. Sound assumptions in terms of target-related efficacy and safety increase the confidence (and possibly reduce the risk of failure) for their modulators.

Recombinant DNA[15] and polymerase chain reaction (PCR)[16] technologies allowed gene-driven *target identification (TI)*. Initially, targets were identified through *functional cloning*[17] (Fig. 1.3, left). Disease-related biochemical abnormalities of proteins prompted researchers to isolate and characterize the altered proteins. The structure of their encoding genes and their chromosomal location were determined. Sickle cell anemia and beta hemoglobin/HBB[18] are examples of TI via functional cloning.

Later, *positional cloning*[19] required the approximate chromosomal position of putative targets, identified by scanning up to 5 millions base pairs for disease-related mutations. Positional cloning-based TI relied upon chromosomal rearrangements (e.g., chronic myelogenous leukemia/CML-breakpoint cluster region-Abelson/BCR-ABL[20]), including extended trinucleotide repeats (e.g., Machado-Joseph disease/MJD-ataxin3/ATXN3[21]) (Fig. 1.3, middle). Even without major DNA rearrangements, positional cloning (e.g., spinal muscular atrophy/SMA-survival motor neuron/SMN gene[22]), or a hybrid positional/functional approach[18] (e.g., Marfan syndrome-fibrillin/FBN[23]) were successful in TI by mid-late '90s.

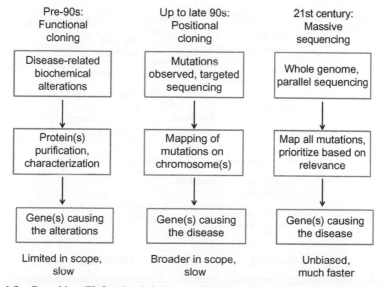

Pre-90s: Functional cloning	Up to late 90s: Positional cloning	21st century: Massive sequencing
Disease-related biochemical alterations	Mutations observed, targeted sequencing	Whole genome, parallel sequencing
Protein(s) purification, characterization	Mapping of mutations on chromosome(s)	Map all mutations, prioritize based on relevance
Gene(s) causing the alterations	Gene(s) causing the disease	Gene(s) causing the disease
Limited in scope, slow	Broader in scope, slow	Unbiased, much faster

FIG. 1.3 Gene-driven TI: functional cloning, positional cloning, and massive sequencing.

Technology improvements led to *massive parallel DNA sequencing*.[24] Times-to-genomes were reduced from ≈40 weeks (one disease-related mutation *per* year *per* lab) to ≈2 weeks by cutting-edge labs[25] (Fig. 1.3, right). The *-omics revolution*, entailing gene/genome, RNA/transcriptome and protein/proteome switches, produced an enormous flow of data (although handling the expectations of scientists and patients was not easy[26]). Primary high-throughput (HT) sequencing data were used in genome-wide association studies (GWAS)[27] to identify disease-associated variants as therapeutic options.

HT experimentation, data mining and computing, statistics and mathematical modeling defined complex biological systems through *systems biology* (Fig. 1.4).[28]

HT-driven comparative genomics (species-specific[29] and diseased vs healthy[30] gene expression profiling), epigenomics (epigenetic DNA modifications[31]), transcriptomics (RNA expression profiling[32]), and functional proteomics (functional protein expression profiling[33]) provided information. Data were analyzed in silico to establish networks/connections among biological entities, and to elucidate their interactions. Models of complex biological systems were built, and their behavior (i.e., functions, regulations) was studied.[28]

Systems biology was applied to physio- and pathological pathways, and contributed to TI against multifactorial diseases.[34] Systems biology analysis merged GWAS and prior knowledge into genome-wide regulatory networks (GWRN),[35] identifying key regulators of healthy and diseased processes.

Many drug targets belong to complex networks (Fig. 1.5), where multiple interactions lead to biological functions.[36] Networks include nodes (mostly proteins, but also RNA and DNA molecules) and edges connecting them (either

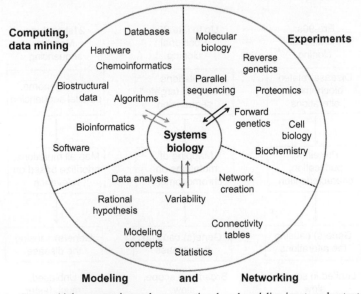

FIG. 1.4 Systems biology: experimental, computational, and modeling inputs and outputs.

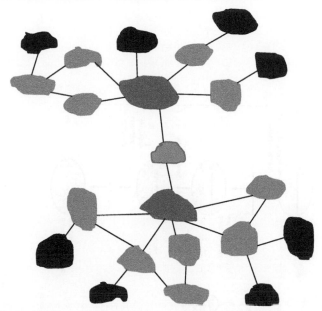

FIG. 1.5 Systems biology networks: central nodes, intermediate nodes, peripheral nodes, and edges.

a physical interaction, or a statistical correlation among two connected nodes).[37] Nodes are connected in accordance with empirical evidence—they are modified once novel information is available. Comparison of the same network in healthy and diseased samples enabled the identification of disease-influencing nodes, i.e., putative therapeutic targets.[38]

The connectivity of a node influences its selection as a drug target.[37] Modulation of central nodes/hubs (two gateways to networks, dark, Fig. 1.5) could cause significant effects on the whole network, but could also lead to unwanted side effects. Modulation of peripheral nodes (black extremities) should not cause toxicity, but could be insufficient to revert a diseased phenotype. Intermediate nodes (light color) are a compromise that often qualified as drug targets.[39] Bridging nodes (single connection among networks, Fig. 1.5) connecting independent networks could regulate network-network interactions (often altered in disease conditions) without affecting each network (avoiding putative toxicity), thus being putative good targets.

Network analysis elucidated pathological processes and identified putative targets. Examples include deregulated pathways and putative disease genes in glioblastoma[40]; key genes to influence cancer metastasis from breast to lungs[41]; differentially regulated proteins during HCV infection[42]; and an angiogenesis network preventing the transition of tumors from dormant to malignant states.[43]

TI benefited from *RNA interference (RNAi)*,[44] i.e., gene silencing through exogenous RNA-driven destruction, or suppressed translation of messenger RNAs (mRNAs). Endogenous RNAi processes inspired selective RNAi methods based on exogenous RNAs. Both processes (Fig. 1.6,

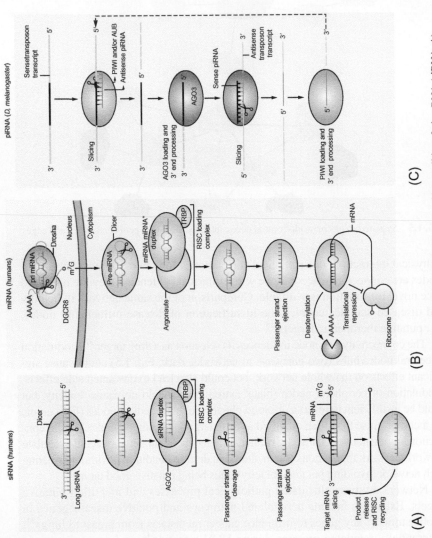

FIG. 1.6 RNA interference (RNAi) mediated by microRNA (miRNA, B) and by small interfering RNA (siRNA, A).

left and middle) are described here, while PIWI-interacting RNAs (piR-NAs,[45] Fig. 1.6C) are not dealt with.

Mammalian gene expression is regulated by *microRNA* (*miRNA*) non-coding genes (Fig. 1.6B).[46] Hairpin-shaped primary miRNA transcripts (pri-miRNA) are processed to precursor miRNA (pre-miRNA, ≈70 nucleotides/nt) in the nucleus by the RNA hydrolase (RNAse) Drosha.[47] Pre-miRNA are cleaved to mature miRNas (double-stranded, 2 nt-extended active sequences, ≈21–25 nt) in the cytoplasm by the RNAse Dicer.[48] miRNAs are composed of a guide miRNA strand and a passenger miRNA* strand. The strands contain limited mismatches, mostly in the central region (nt 9–11).[49] Mature miRNAs are loaded onto the RNA-induced silencing complex (RISC).[50] RISC contains Argonaute endonucleases[51] and Argonaute-bound Gly/Trp repeat-containing proteins of 182 kDa (GW182[52]). miRNA:miRNA* mismatches prevent cleavage by Argonaute. Rather, cleavage-independent miRNA unwinding leads to the recognition of the single-stranded (ss) guide by its target mRNA.[50] The passenger strand is lost and degraded by RISC proteins.[53]

miRNAs do not cause target mRNA cleavage. Their effects on target mRNAs include miRNA-mediated deadenylation, decapping, and degradation,[54] translational repression through impaired ribosomal recruitment,[55] and combinations of the two[56] (Fig. 1.6B).

Gene silencing by exogenous RNA sequences in *C. elegans*[57] suggested *small interfering RNA* (*siRNA*) as a tool in early drug discovery.[58] RNA libraries can be delivered to model organisms (*C. elegans*,[59] *Drosophila*[60]), or to mammalian cells.[61] Long dsRNA (processed by the Dicer/RISC machinery),[62] short ds siRNA (loaded onto the RISC complex),[63] or vector-based small hairpin RNA (shRNA, high efficiency in gene silencing)[64] can be used as exogenous RNA.

RISC processing of siRNAs is shown in Fig. 1.6A. Dicer cleavage and RISC loading of complementary ds siRNA strands enable passenger strand cleavage by Argonaute/AGO2.[65] The passenger strand is released from RISC/AGO2 and degraded, while the ss guide strand binds to its target mRNA. Complementarity between the guide strand and target mRNA ensures cleavage of the latter by Argonaute[66] and release from RISC (Fig. 1.6A). Gene silencing by exogenous siRNAs (via Argonaute-driven mRNA slicing and degradation), or by endogenous miRNAs (via mRNA deadenylation or impaired ribosomal recruitment) differs in mechanism but leads to similar results.

Exogenous RNAs were used for transfection,[67] although synthetic transfection reagents[68] improve their efficiency. Genome-wide RNAi libraries (ds-, si-, or sh-RNAs)[69] targeted the annotated genes of whole organisms, as unbiased search pools for TI. RNAi sublibraries focused against target classes (i.e., kinases,[70] G-coupled protein receptors/GPCRs,[71] transcription factors,[72] the druggable genome[73]) were also introduced. Off-target effects and drawbacks of RNAi were observed,[74] but its usefulness is undisputed. RNAi-based HT- (HTS) or medium-throughput screening (MTS) campaigns identified, *inter alia*, disease-associated genes in damage survival pathways,[75] signal transduction,[76]

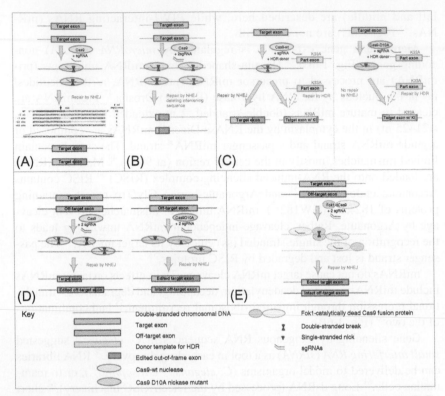

FIG. 1.7 CRISPR-Cas9 gene editing: process and main components (A), recent improvements (B–E).

cellular import/export,[77] organelle formation,[78] cell division,[79] and neuronal protein aggregation.[80]

Genome-wide gene editing by the bacterial *RNA-guided clustered regularly interspaced short palindromic repeats (CRISPR)–Cas9 nuclease*[81,82] was recently introduced as an efficient TI method (Fig. 1.7, quadrant (a)).

Cas9 is expressed in mammalian cells, then treated with a library of single-guide RNA sequences (sgRNAs). An sgRNA directs the nuclease to a complementary dsDNA region (target exon in Fig. 1.7) that is cleaved by Cas9 with high specificity. The dsDNA break is misrepaired by nonhomologous end joining (NHEJ, Fig. 1.7), a nonspecific endogenous pathway. The misrepaired target gene is inactive, and its inactivation modulates a therapeutically relevant phenotype.[83] Improvements in CRISPR-Cas9 editing include using two sgRNAs targeting close regions of a gene[84] (quadrant (b)); a single-point CRISPR mutation[85,86] (quadrants (c) and (d)); and a fusion protein between catalytically inactive Cas9 and the *Flavobacterium okeanokoites* I (*Fok*I) endonuclease (fCas,[87] quadrant (e)), Fig. 1.7). High cleavage efficiency and selectivity, with complete avoidance of off-target gene editing, was ensured in genome-wide screenings related to oncology[88] and to toxin-mediated infection.[89]

DNA- and RNA-based, genome-wide TI efforts are effective complementary approaches.[90] A molecular target identified using DNA- or RNA-based methods could resist modulation by small molecules, or even by biological reagents. Chemistry-rich approaches in TI, described next, overcome this limitation.

1.2. CHEMISTRY IN MODERN TARGET IDENTIFICATION: PHENOTYPIC SCREENING, FORWARD CHEMICAL GENETICS

Genome-wide methods in TI were not conceivable in the past. A few abundant proteins could be purified and used to test a limited number of compounds as putative protein target modulators. In particular, natural products (NPs) were useful to identify proteins, pathways, and mechanisms with therapeutic potential.[91] NPs influenced pathological phenotypes—e.g., growth inhibition of bacterial pathogens, cytotoxicity against a tumor cell line. NP probes contributed to the identification of putative therapeutic mechanisms, as described later.

Synthetic and naturally occurring drugs were often used without knowing their target/MoA. Aspirin was identified in early 19th century as the active anti-inflammatory agent in willow bark, while its inhibition of prostaglandin synthesis was reported ≈150 years later.[92] The discovery of an NP and its mechanistic characterization were typically years apart in 20th century. The Investigational New Drug (IND) application for Phase I testing of a clinical candidate does not require a clarified MoA,[93] and a known MoA is not considered a clinical advantage.[14] ≈8% of drugs approved between 2000 and 2012 did not have a confirmed MoA.[94]

Forty five out of seventy seven first-in-class, innovative small-molecule drugs approved by FDA between 1999 and 2013[95] were discovered using a *target-based approach*—the R&D project was aimed against a molecular target, and rational drug design was used to identify and optimize hits, leads, and candidates. Thirty two small-molecule drugs came from *phenotype-based approaches*—their precursors were identified through a phenotypic screening, and their MoA was later elucidated.

The phenotypic drug set is smaller because of strategic decisions by big pharma companies. Target-based compounds approved in the last 15 years came from the switch from pre-1980, phenotypic-driven to HT-technology-dependent, target-driven drug discovery.[96] One could argue for[95] or against[97] target-based/rational drug discovery, but the number of marketed drugs per year has not increased despite an increase in costs and timelines. Thus, target-based drug discovery and even the "omics revolution" did not translate (yet) into more high-quality drugs.

Let us consider diseases where known molecular targets are few, if any. Neurodegenerative diseases (NDs), such as Alzheimer's disease (AD), exemplify the difficulties of target-based drug discovery.[98] Disease-modifying targets against NDs are largely unknown—their identification and validation would enable the market launch of first-in-class, effective ND-targeted drugs. ND-recapitulating models, based on cellular or animal screens, are questionable. Contradictory effects of leads (beneficial in early-disease stages and detrimental in late stages, or vice versa) were observed, but not rationalized yet. Difficulties in monitoring the development and the modulation of ND-related phenotypes (lack of molecular markers) prevented the identification of druggable molecular targets against NDs.

Efforts against NDs and AD targeted many putative molecular targets and pathways in the past 30 years.[99] Chemical and biological candidates entered clinical development, but only a handful of drugs—mostly acting on symptoms at an early ND stage—were marketed.[100] A validated pool of disease-associated targets whose modulation would lead to disease-modifying outcomes is thus needed.

Phenotypic screening—if the phenotype is truly relevant to the inception and/or progression of one or more NDs—provides two priceless outcomes if hits are found.[101] A novel disease-modifying molecular target is identified and can be modulated—i.e., it is druggable.

The novel target is identified in an unbiased manner from a (complex) phenotype recapitulating an ND. If the phenotype-causing pathway is not characterized, hits contribute to elucidate a pathological mechanism. As to hits, this is where genomics-, RNA-, and chemistry-based phenotypic screening differ. System biology may identify a putative gene target via data mining, or through HT genome sequencing and identification of disease-connected mutations (path 1, Fig. 1.8).

Disease states are observed in silico (data-mining-mutated genes) or in vitro (genome-sequencing-mutated genes); are recapitulated in vitro (mutated-gene-containing cell lines) or in vivo (mutated-gene-containing zebrafish, *Drosophila*, rodents); and are irreversibly modified by genomic interventions (gene ablation in cell lines and whole organisms). A gene-disease connection is established, but a "hard"/nondrug-like modulation of the phenotype does not prove the druggability/putative therapeutic modulation of the target (path 1, Fig. 1.8).

RNAi and CRISPR-induced gene editing have therapeutic implications (path 2, Fig. 1.8). Cellular- and whole-organism-recapitulating disease phenotypes can be reversibly modulated via RNAi (mRNA/translation inhibition), or irreversibly via gene editing (gene cleavage-misrepair-inactivation). In both cases, a gene-disease connection is established. Identified siRNAs and misrepaired/edited genes could be translated into RNAi-based therapeutics[44] and CRISPR-based gene therapies[102] -both tested up to preclinical studies[103,104] (path 2, Fig. 1.8). Clinical trials on RNAi-based therapeutics[105] took advantage of synthetic delivery agents.[68] More data—especially regarding CRISPR gene editing—are needed to evaluate their potential and their drawbacks in therapy.[76,102]

Pathological phenotypes in cellular systems or model organisms can be modulated by small molecules in a reversible, time-dependent manner (Fig. 1.9).

A library of drug-like, diverse small molecules can be synthesized, or purchased (step 1, Fig. 1.9). The collection size varies between ≈1000 (medium,

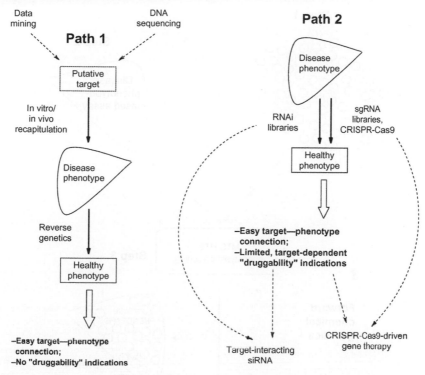

FIG. 1.8   Biology-driven TI: systems biology (path 1), RNAi and gene editing (path 2).

MTS campaign) and >100,000 compounds (large, HTS campaign). In parallel, a disease-connected phenotypic assay is set up (step 2). Each library member is tested in an MTS or HTS phenotypic screening (step 3). Most compounds do not induce phenotypic changes (step 4'). A few positives cause a phenotypic reversion to a nondiseased state (step 4). The phenotypic switch is reversible, and experimental conditions (compound concentration, exposure time, washout period) can be adjusted to maximize the effect of positives. After assay reconfirmation, validated hits (dots, rescued phenotype, Fig. 1.9) are identified.

The small-molecule modulator-molecular target connection (covered in Chapter 2) is harder to determine than detecting a mutation, identifying an RNAi-interacting gene, or identifying a misrepaired gene.[106] Conversely, small-molecule modulators prove that a disease-connected pathway can be targeted by drug-like compounds, and is druggable.[107] The use of drug-like chemical diversity to identify and validate novel molecular targets using phenotypic screening was named *forward chemical genetics (FCG)* by Schreiber in 1998[108] (Fig. 1.9).

Out of 32 small-molecule drugs from phenotype-based approaches, approved by FDA between 1999 and 2013,[95] 25 were either NPs, or semisynthetic NP derivatives. The identification of their molecular target was a biased chemocentric process[95] that did not require the phenotypic screening of a chemical

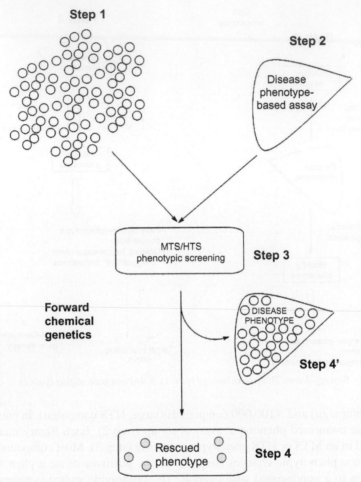

FIG. 1.9 Phenotypic screening-based forward chemical genetics (FCG): the process and its main components.

library. The remaining seven FDA-approved drugs resulted from an R&D project initiated with *FCG* and phenotypic screening.

An example of FCG in TI is presented in the next section, while key implements of any FCG approach will be described elsewhere (chemical tools and probes—Chapter 2; cell-based phenotypic assays—Chapter 3).

1.3. KINESIN SPINDLE PROTEIN (KSP, Eg5)—MONASTROL

The cell division cycle of animal cells[109] is depicted in Fig. 1.10.

A long *interphase* process starts when cells grow in size and replicate their organelles (*gap/growth phase 1*, **G1**, Fig. 1.10).[110] When G1 is completed (*G1/S*

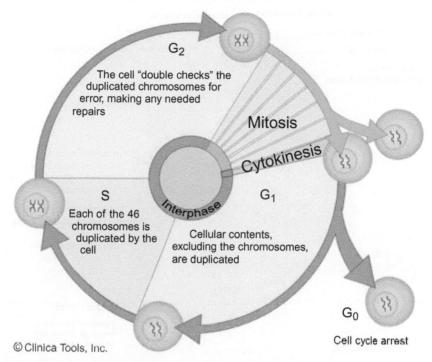

© Clinica Tools, Inc.

FIG. 1.10 The mammalian cell cycle: main phases and checkpoints.

checkpoint, restriction point[111]), DNA replication takes place (*synthesis/S* phase,[112]) until duplication of chromosomal DNA in two sister chromatids, and duplication of centrosomes (microtubule/MT-organizing centers/MTOCs made by centrioles surrounded by unorganized MT precursor/connected proteins).[113] Sister chromatids are kept together by cohesin, a protein that binds and forms a ring around chromosomal DNA before the completion of its duplication.[114] Cell growth and protein synthesis continue in the **G2** phase.[115] The transition of interphase to mitosis (Fig. 1.10) is controlled by the *G₂/M checkpoint*,[116] ensuring with the earlier G_1/S checkpoint that only undamaged cells are replicated.

Mitosis/M[117] is a five-stage process, in which karyogenesis (nuclear division) takes place and two twin sets of chromosomes are lodged in two nuclei. Its components are shown in Fig. 1.11.

After G_2 completion (**G2** section, Fig. 1.11), prophase entails condensation of loosely packed DNA-protein complexes (chromatin) into X-shaped sister chromatid pairs joint at midsection by centromeres in the nucleus,[118] disappearance of the nucleolus,[119] nuclear membrane weakening and stoppage of gene transcription.[120] Two fiber-sprouting centrosomes associated onto the nucleus membrane[121] become visible in prophase.

Prometaphase starts with phosphorylation of nuclear lamins by the mitosis-promoting factor/MPT, causing dissolution of the nucleus into small vesicles.[122]

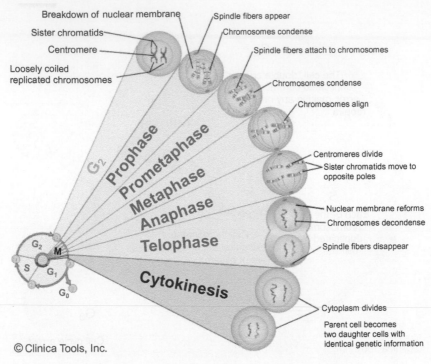

Breakdown of nuclear membrane
Sister chromatids
Centromere
Loosely coiled
replicated chromosomes

Spindle fibers appear
Chromosomes condense
Spindle fibers attach to chromosomes
Chromosomes condense
Chromosomes align

Centromeres divide
Sister chromatids move to
opposite poles

Nuclear membrane reforms
Chromosomes decondense

Spindle fibers disappear

Cytoplasm divides
Parent cell becomes
two daughter cells with
identical genetic information

Prophase
Prometaphase
Metaphase
Anaphase
Telophase
Cytokinesis

G_2

G_2 M
S G_1
G_0

© Clinica Tools, Inc.

FIG. 1.11 Mammalian mitosis: main stages, cellular processes, and organelles.

Two kinetochores[123] (proteinaceous MT-binding structures on each sister chromatid close to the centromere) are formed, two MT types sprout from each centrosome and invade the nuclear space. Kinetochore MTs[124] join kinetochores, connecting the centrosomes via multiple kinetochore MT-kinetochore interaction. Polar MTs,[125] sprouting from a centrosome, laterally interact with polar MTs from the other centrosome, establishing the mitotic spindle.[126] Astral MTs[127] connect centrosomes to the cell envelope through cell-membrane-anchored, (−) end-directed ATP dynein motors, and contribute to the positioning and orientation of the mitotic spindle.[128]

The transition from prometaphase to *metaphase* is completed when all kinetochores are connected to kinetochore MTs, and each chromosome is aligned with the metaphase/equatorial plane.[129] The alignment of chromosome pairs is ensured by equipotent pulling of each kinetochore MT couple toward its centrosome.[130] Even a single misaligned chromosome pair could trigger the spindle checkpoint and stop mitosis.[131]

Anaphase starts with cohesin cleavage by the anaphase-promoting complex, releasing two daughter chromosomes.[132] Kinetochore MTs shorten, due to MT disassembly from the kinetochore end, pulling the V-shaped daughter chromosomes toward each centrosome (*anaphase A*).[133] Polar MTs from different centrosomes pull away from each other through (+) end-directed, ATP kinesin motor-dependent mechanisms. Cells elongate, and the mitotic spindle collapses (*anaphase B*).[134] *Telophase* completes the mitotic process. Two nuclei are assembled by dephosphorylation of nuclear lamins in membrane vesicles.[135] Newly formed nuclei enclose the chromosomes, but not the centrosomes. The nucleolus reappears, spindle fibers uncoil from chromosomes and degenerate, chromosomes unpack and resume the interphase-like extended chromatin appearance.[136]

Cell division is completed by *cytokinesis* (Figs. 1.10 and 1.11). A contractile proteinaceous ring, made by nonmuscle myosin II and actin, forms at the equatorial cell plane.[137] The ring moves inwards, due to ATP-driven myosin movement along actin filaments, to generate a midbody section that cleaves the cell in two daughter cells.[138] Once cytokinesis is completed, cells exit the cell cycle and enter the *resting*/G0 phase (Fig. 1.10).

Up to hundreds of proteins constitute organelles (i.e., tubulin, actin), molecular motors (i.e., kinesins, dynein), cell-cycle regulators (i.e., cyclin-dependent kinases/CDKs), cell-cycle inhibitors (i.e., p53 and p27), and transcription factors (i.e., signal transducers and activators of translation/STATs, NF-κB). Some modulate dysregulation of the cell cycle, such as hyperproliferation in cancer. One should select relevant (major contribution to dysregulated cell cycle) and druggable targets (modulation impacting the pathological phenotype) devoid of off-target effects.

Several mitotic MoAs are clinically validated, due to known drugs acting on mitosis (either selectively, or acting also on interphase and cytokinesis). *Microtubule-targeting agents (MTAs)*[139] bind to tubulin and/or to MTs, and alter their stability during mitosis. Inhibitors of tubulin polymerization / *MT destabilizers* (such as colchicine **1.1**[140] and vinblastine **1.2**,[141] Fig. 1.12) and inhibitors of tubulin depolymerization/*MT stabilizers* (such as taxol **1.3**, Fig. 1.12[142]) are marketed anticancer drugs. They are NPs with an unknown MoA used to identify and characterize novel therapeutic avenues.

Agents targeting *cell-cycle-regulating kinases*[143] and *molecular motors*[144] prevent the replication of cancer cells and control tumor growth. The former cause an abnormal exit from the cell-cycle process, the latter indirectly impair MT assembly and disassembly. One of them was approved for cancer treatment,[145] and the clinical development of others is ongoing. Aurora kinase inhibitors[146] such as alisertib (MLN8237, **1.4**, Fig. 1.13) cause premature mitotic exit for replicating cancer cells due to chromosomal misalignment and incomplete kinetochore-kinetochore MT attachment.

Polo-like kinase (PLK) inhibitors[147] such as volasertib (BI-6727, **1.5**) inhibit cytokinesis, and affect the formation and the functionality of the bipolar spindle. CDK4 and CDK6-specific inhibitors[148] such as palbociclib (Ibrance®, PD0332991, **1.6**, Fig. 1.13) impair the regulation of the G_1/S checkpoint, and of other checkpoints needed by cancer cells to complete replication.

In 1999, only MTAs were clinically validated, although their binding modes to tubulin and MTs were not yet known.[149] Novel molecular targets against dysregulated cell cycles in oncology were actively pursued. A Harvard team

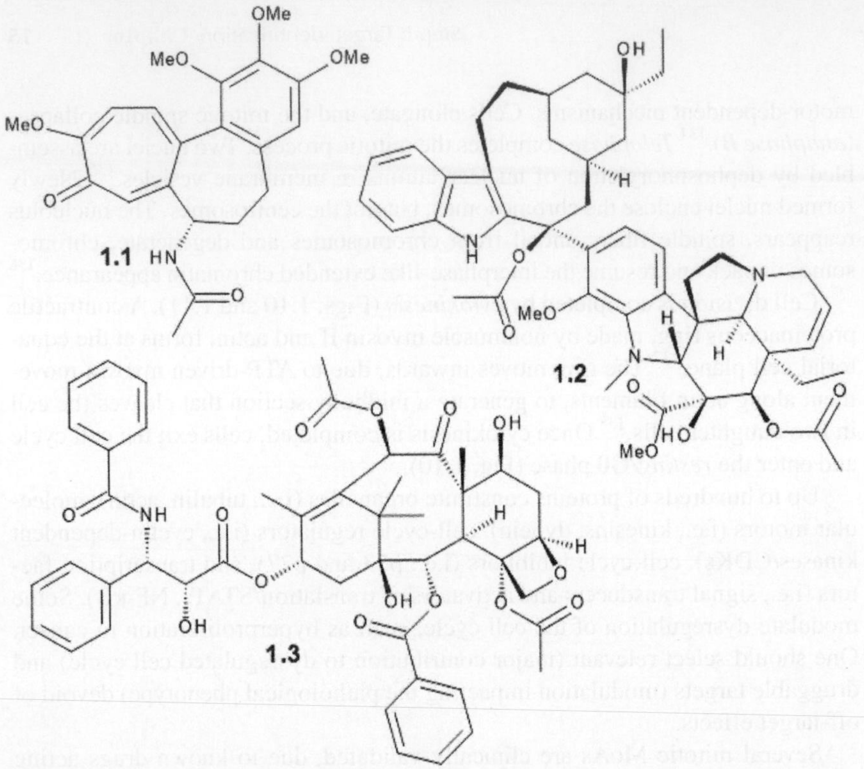

FIG. 1.12 Structure of microtubule-targeting agents (MTAs): colchicine **1.1**, vinblastine **1.2**, taxol **1.3**.

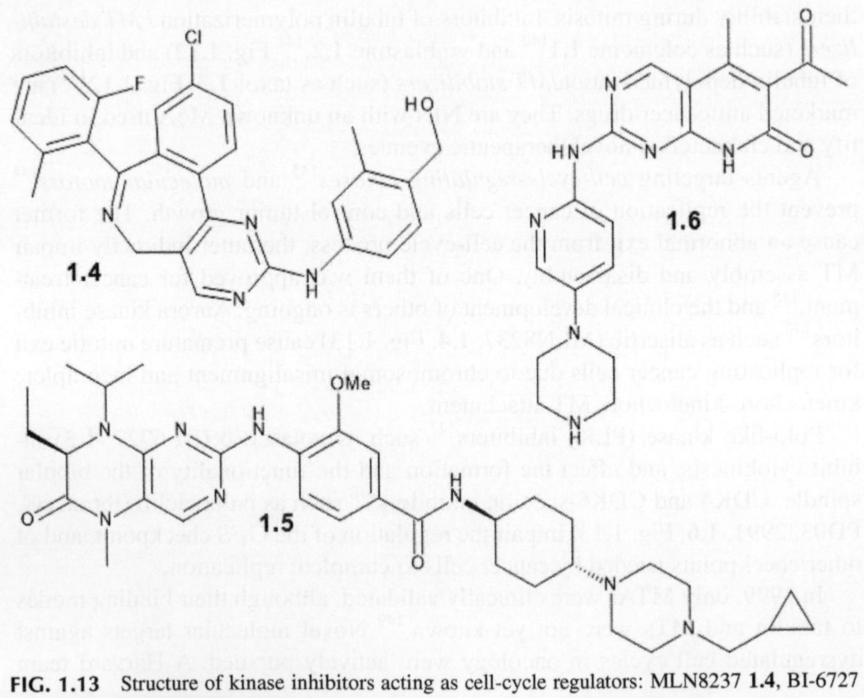

FIG. 1.13 Structure of kinase inhibitors acting as cell-cycle regulators: MLN8237 **1.4**, BI-6727 **1.5**, Ibrance®-PD0332991, **1.6**.

reported then a phenotypic/FCG screening aimed toward mitosis-affecting small molecules.[150,151] The screening strategy is depicted in Fig. 1.14.

The primary HTS assay was based on the monoclonal antibody TG-3,[152] recognizing a mitosis-specific phosphorylated form of the nucleolar protein nucleolin.[153] A cell-based cytoblot assay format[154] was used on A549 lung carcinoma cells to detect post-translational modifications (PTMs) on proteins. A commercial, diversity-based library (16,320 compounds[155]) was tested at 45 µM for 18 h (step 1, Fig. 1.14). One hundred and thirty nine positives (≈0.85%) induced a >2.5 fold excess of phosphorylated nucleolin, i.e., caused the blockage of cells in mitosis.

To exclude MTAs from the positives, a secondary tubulin polymerization-based assay (step 2)[156] entailed testing positives (50-µM concentration) in presence of fluorescently labeled tubulin.[157] Fifty two compounds (group I)

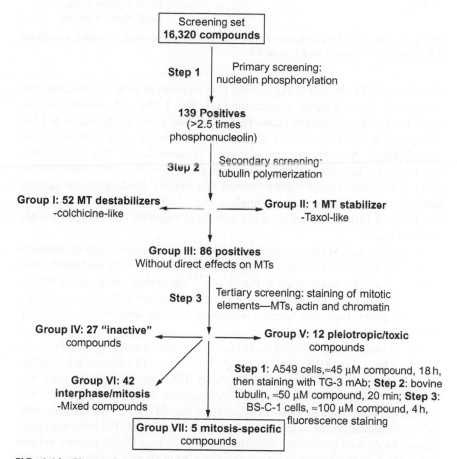

FIG. 1.14 Phenotypic screening-centered strategy toward novel mitotic targets: the process, chemical diversity, and profiling assays.

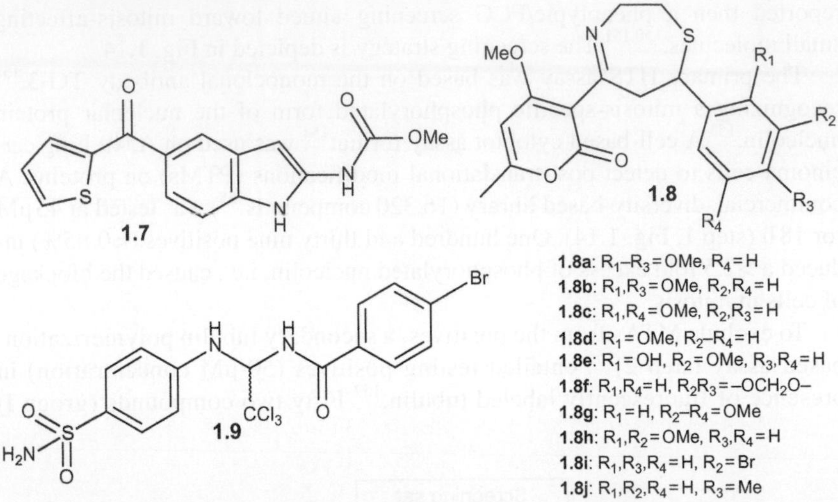

1.7

1.8

1.8a: R_1–R_3=OMe, R_4=H
1.8b: R_1,R_3=OMe, R_2,R_4=H
1.8c: R_1,R_4=OMe, R_2,R_3=H
1.8d: R_1=OMe, R_2–R_4=H
1.8e: R_1=OH, R_2=OMe, R_3,R_4=H
1.8f: R_1,R_4=H, R_2,R_3=–OCH$_2$O–
1.8g: R_1=H, R_2–R_4=OMe
1.8h: R_1,R_2=OMe, R_3,R_4=H
1.8i: R_1,R_3,R_4=H, R_2=Br
1.8j: R_1,R_2,R_4=H, R_3=Me

1.9

FIG. 1.15 Structure of MTAs identified in the mitosis-targeted phenotypic screening: nocodazole **1.7**, tetrahydrothiazepines **1.8a–f**, synstab **1.9**.

destabilized MTs by inhibiting tubulin polymerization with a colchicine-like mechanism. Among them, nocodazole (**1.7**, Fig. 1.15), and several disubstituted tetrahydrothiazepines (**1.8a–f**). Such MTAs either destabilized MTs in interphase and mitosis (e.g., **1.8b**), or showed weaker effects in interphase (e.g., **1.8e**).[151] A single compound (group II, synstab, **1.9**, Fig. 1.15) stabilized MTs by inhibiting tubulin depolymerization with a taxol-like mechanism. Competition experiments showed that the MT binding site of synstab partially overlaps with the taxol-binding site.[151] Group I/II compounds were not pursued further, although could represent valuable chemotypes toward new MTAs.[158]

Eighty six non-MTA positives (group III) were incubated with mammalian epithelial kidney BS-C-1 cells. Fixed cells were stained with immunofluorescence in a tertiary assay (step 3, Fig. 1.14),[159] observing variations in mitotic and interphase elements. The effect of each positive was fully determined in a low-throughput, high-cost format compatible with the small number of group III compounds.

Seven compounds behaved as MTAs with lower potency (group IV, compounds **1.8g–j**, Fig. 1.15). Twenty compounds (group IV) did not affect either the appearance of mitotic/interphase elements (DNA, mitotic spindle, actin filaments), or their dynamic distribution.[150] An increase of normal mitotic cells was confirmed, possibly due to weak phenotype-less MTAs, to interactions with cell-cycle regulators, or to labile interactions in fixed cells. Twelve compounds (group V) showed pleiotropic effects reconducible to multiple actions, related to mitotic disturbance and apoptosis in cells. Group IV and V compounds were not pursued further.

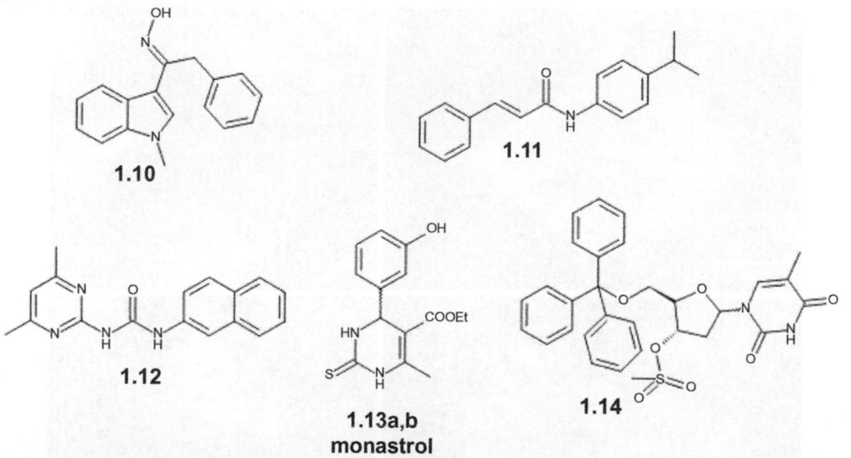

FIG. 1.16 Structure of MTAs identified in the mitosis-targeted phenotypic screening: compounds 1.10–1.14.

Forty two positives (group VI) affected MTs (depolymerization, abnormal spindle formation) and/or chromosomes (misalignments). Concerns about multiple targets/polypharmacology suggested that TI-driven efforts on them could have been difficult, putting group VI compounds on hold. The focus shifted on the remaining five mitosis-specific compounds (group VII, **1.10–1.14**, Fig. 1.16). They did not cause any effect in interphase cells, while they affected chromosomes and MTs during mitosis.

Compound **1.10** formed a double spindle and caused partial chromosome misalignment, while **1.11** caused chromosomal misalignment without affecting the bipolar spindle.[151] Compounds **1.13a,b** and **1.14** caused the same alteration, with the former mixture being more potent. The bipolar spindle observed in untreated BS-C-1 cells (Fig. 1.17A) was replaced after 4 h treatment by a mono-astral structure in >90% of the cells (Fig. 1.17C). Sister chromatids (light blue) were anchored to kinetochore MTs, stood at the end of the "rays," and showed the oscillatory behavior of MT-chromosome constructs linked to a single pole/centrosome.[160] The distribution of chromatin and MTs during interphase was not altered (compare b-untreated and d-treated cells with **1.13a,b**, Fig. 1.17), confirming its mitosis-specific MoA.

Due to the induced monoastral phenotype, the racemic mixture **1.13a,b** was named monastrol and was selected for TI efforts.

The specific effects of monastrol were due to the modulation of a single, unknown molecular target. As said earlier, the phenotype-target connection in FCG is usually more complex than genome- or RNA-based TI—see next chapter. That was not the case for monastrol, as a monoastral phenotype was earlier observed after genetic/biological challenges in *Drosophila* when the kinesin KLP61 gene was mutated.[161] Similar phenotypes—block of mitotic spindle formation, lack of spindle pole separation—were observed in fungi when

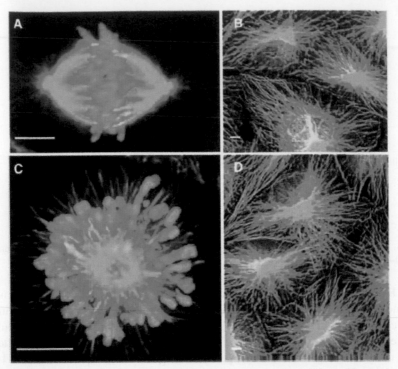

FIG. 1.17   Mitotic spindle in untreated BS-C-1 cells (top, A–B); monoastral structure in BS-C-1 cells treated with monastrol (68 μM, bottom, C–D). α-tubulin, *green*; chromatin, *blue*.

KLP61-homologue kinesins bimC and cut7 were inactivated.[162,163] *Xenopus* egg extracts treated with a monoclonal antibody (mAb) specific for the homologous kinesin Eg5 showed monoastral mitotic phenotypes.[164]

The same phenotype was observed after treatment of human HeLa cells with a human Eg5-directed mAb.[165] Human Eg5 (known as kinesin-5, kinesin spindle protein/KSP and kinesin family member 11/KIF11) is an ATP-dependent kinesin homotetramer.[166] Eg5 connects two oppositely oriented MTs in prophase and prometaphase, and promotes the migration of centrosomes in opposite directions during the formation of the mitotic spindle.[167]

Due to the monoastral phenotype observed by inactivation of Eg5, monastrol was tested for its inhibition of human Eg5 in a cell-free assay.[168] Monastrol inhibited Eg5-driven MT motility with a 14-μM IC_{50}, similar to the 22-μM IC_{50} observed in the cytoblot/FCG assay.[150] The effect of monastrol was reversible, as its removal led to the restoration of MT velocity after washout, and to completion of mitosis in 4 h.[150] The effect of monastrol was specific, as it did not affect MT motility driven by conventional kinesin, and did not alter localization and organization of Golgi reticulum and lysosomes in interphase cells up to 200-μM concentrations.[150]

   The identification of monastrol and of its target—Eg5—allowed the characterization of the physio- and pathological functions during the cell cycle of Eg5, and the elucidation of its binding mode with monastrol. Monastrol prevented centrosome separation after duplication in a dose-dependent manner (centrosome colocalization at 100-μM monastrol),[169] recapitulating the phenotype induced by genetic inactivation of Eg5.[161] Monastrol treatment did not influence localization of Eg5 on MTs, did not inhibit MT-Eg5 binding, and did not promote disassembly of preformed bipolar spindles in BS-C-1 cells.[167]

   During mitosis, the motor domain of Eg5 binds to MTs (step 1, Fig. 1.18), recruits ATP (step 2), produces energy via ATP hydrolysis (step 3), releases ADP after detachment from MTs (steps 4 and 5), and recruits again released Eg5 onto MTs (step 6, Fig. 1.18).[170]

   Monastrol inhibits basal and MT-stimulated ATPase activity of Eg5 (IC_{50} of 6.1 and 34 μM). Monastrol (in particular its (S)-enantiomer **1.13a**) neither competes with ATP-Eg5 nor with MT-Eg5 binding, but binds to an Eg5 allosteric site. It binds to the Eg5-ADP binary complex (step 4', Fig. 1.18), and with

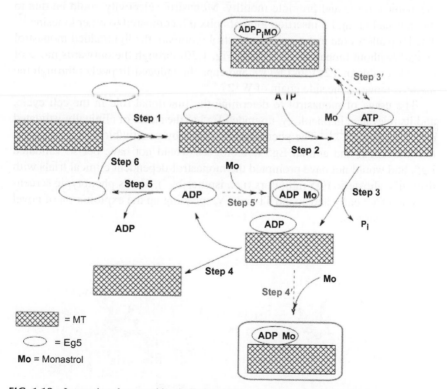

FIG. 1.18 Interactions between kinesin Eg5 and MTs with and without monastrol.

lower affinity to the MT-Eg5-ADP ternary complex (step 5′). The catalytic cycle cannot continue, as ADP cannot be released from either free, or MT-bound Eg5 (Fig. 1.18).[170]

A putative effect on in situ ATP resynthesis suggested the regeneration of ATP from ADP and P_i in the Eg5 active site due to monastrol stabilization of an ATP resynthesis-favoring conformation (step 3′, Fig. 1.18).[171] A loop between ATP hydrolysis and resynthesis, with reduced/impaired P_i and ADP release, would also have impaired the Eg5 catalytic cycle (Fig. 1.18).[171]

A druggable target is even more relevant for drug discovery, when its binding with a small-molecule modulator is structurally characterized. The structure of a ternary complex between Eg5, Mg^{2+}·ADP, and monastrol showed the location of the allosteric binding site. It consists of an induced-fit pocket close (≈ 12 Å) to the ATP/ADP binding site, contoured by 20 aminoacids between helix a3 and the insertion loop (L5) of helix α2 (Fig. 1.19).[172]

Monastrol binding does not involve the nucleotide pocket, induces the formation of an allosteric rigid site, modifies the geometry of the MT binding site on Eg5, and decreases its flexibility. It possibly impairs its aptitude to generate mechanical force and promote motility. Monastrol selectivity could be due to the unusually long L5 insertion loop of helix α2, compared to other kinesins.[173] The loop allows the insertion of monastrol (compare the light/added monastrol vs dark/without monastrol structures, Fig. 1.20) through the outwards move of the side chains of Y211 and R119, and caps the induced-fit pocket through the inwards move of the side chain of W127.[172]

The usage of monastrol to determine the functional role in the cell cycle, and its relevance in oncology, upgraded Eg5 to the status of clinically validated oncology target.[144,174] The phenotype observed by genetic/biological intervention on fungal and animal Eg5 homologues would not have per se validated Eg5, and would not have prompted the monastrol-dependent clinical trials with drug-like allosteric Eg5 inhibitors (i.e., ispinesib[175]). That's phenotypic screening and FCG: early and sound TI and TV, speeding up the exploitation of novel molecular targets.

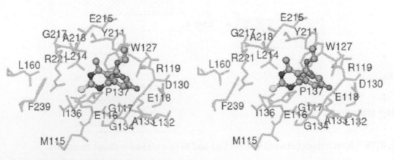

FIG. 1.19 Structure of the allosteric binding site of Eg5 complexed with monastrol.

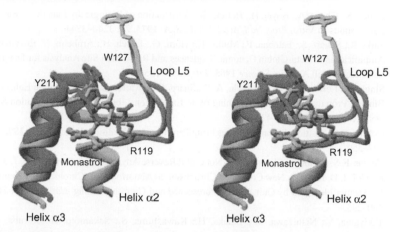

FIG. 1.20 Structure of the allosteric binding site of Eg5 before *(dark)* and after *(light)* monastrol binding.

REFERENCES

1. Hopkins, A. L.; Groom, C. R. The Druggable Genome. *Nat. Rev. Drug Discov.* **2002**, *1*, 727–730.
2. Claverie, J.-M. What If There Are Only 30,000 Human Genes? *Science* **2001**, *291*, 1255–1257.
3. Walke, D. W.; Han, C.; Shaw, J.; Wann, E.; Zambrowicz, B.; Sands, A. In Vivo Drug Target Discovery: Identifying the Best Targets From the Genome. *Curr. Opin. Biotechnol.* **2001**, *12*, 626–631.
4. Pertea, M.; Salzberg, S. L. Between a Chicken and a Grape: Estimating the Number of Human Genes. *Genome Biol.* **2010**, *11*, 206.
5. Arkin, M. R.; Tang, Y.; Wells, J. A. Small-Molecule Inhibitors of Protein-Protein Interactions: Progressing Toward the Reality. *Chem. Biol.* **2014**, *21*, 1102–1114.
6. Rader, R. A. (Re)defining Biopharmaceutical. *Nat. Biotechnol.* **2008**, *26*, 743–751.
7. Rask-Andersen, M.; Masuram, S.; Schioeth, H. B. The Druggable Genome: Evaluation of Drug Targets in Clinical Trials Suggests Major Shifts in Molecular Class and Indication. *Annu. Rev. Pharmacol. Toxicol.* **2014**, *54*, 9–26.
8. Regnier, S. What Is the Value of 'Me-Too' Drugs? *Health Care Manag. Sci.* **2013**, *16*, 300–313.
9. Watson, J. D.; Crick, F. H. C. A Structure for Deoxyribose Nucleic Acid. *Nature* **1953**, *171*, 737–738.
10. Matthaei, H. J.; Jones, O. W.; Martin, R. G.; Nirenberg, M. W. Characteristics and Composition of RNA Coding Units. *Proc. Natl. Acad. Sci. USA* **1962**, *48*, 666–677.
11. Sanger, F.; Nicklen, S.; Coulson, A. R. DNA Sequencing With Chain-Terminating Inhibitors. *Proc. Natl. Acad. Sci. U. S. A.* **1977**, *74*, 5463–5467.
12. http://www.who.int/medicines/areas/traditional/definitions/en/.
13. Mahdi, J. G.; Mahdi, A. J.; Mahdi, A. J.; Bowen, I. D. The Historical Analysis of Aspirin Discovery, Its Relation to the Willow Tree and Antiproliferative and Anticancer Potential. *Cell Prolif.* **2006**, *39*, 147–155.
14. Patel, A. C. Clinical Relevance of Target Identity and Biology: Implications for Drug Discovery and Development. *J. Biomol. Screen.* **2013**, *18*, 1164–1185.

15. Cohen, S.; Chang, A.; Boyer, H.; Helling, R. Construction of Biologically Functional Bacterial Plasmids In Vitro. *Proc. Natl. Acad. Sci. U. S. A.* **1973**, *70*, 3240–3244.

16. Saiki, R.; Scharf, S.; Faloona, F.; Mullis, K.; Horn, G.; Erlich, H.; Arnheim, N. Enzymatic Amplification of Beta-Globin Genomic Sequences and Restriction Site Analysis for Diagnosis of Sickle Cell Anemia. *Science* **1985**, *230*, 1350–1354.

17. Singh, H.; LeBowitz, J. H.; Baldwin, A. S.; Sharp, P. A. Molecular Cloning of an Enhancer Binding Protein: Isolation by Screening of an Expression Library With a Recognition Site DNA. *Cell* **1988**, *52*, 415–423.

18. Collins, F. S. Positional Cloning Moves From Perditional to Traditional. *Nat. Genet.* **1995**, *9*, 347–350.

19. Collins, F. S. Positional Cloning: Let's Not Call It Reverse Anymore. *Nat. Genet.* **1992**, *1*, 3–6.

20. Rowley, J. D. Letter: A New Consistent Chromosomal Abnormality in Chronic Myelogenous Leukaemia Identified by Quinacrine Fluorescence and Giemsa Staining. *Nature* **1973**, *243*, 290–293.

21. Takiyama, Y.; Nishizawa, M.; Tanaka, H.; Kawashima, S.; Sakamoto, H.; Karube, Y.; Shimazaki, H.; Soutome, M.; Endo, K.; Ohta, S.; Kagawa, Y.; Kanazawa, I.; Mizuno, Y.; Yoshida, M.; Yuasa, T.; Horikawa, Y.; Oyanagi, K.; Nagai, H.; Kondo, T.; Inuzuka, T.; Onodera, O.; Tsuji, S. The Gene for Machado-Joseph Disease Maps to Human Chromosome 14q. *Nat. Genet.* **1993**, *4*, 300–304.

22. Lefebvre, S.; Burglen, L.; Reboullet, S.; Clermont, O.; Burlet, P.; Viollet, L.; Benichou, B.; Cruaud, C.; Millasseau, P.; Zeviani, M.; Le Paslier, D.; Frézal, J.; Cohen, D.; Weissenbach, J.; Munnich, A.; Melki, J. Identification and Characterization of a Spinal Muscular Atrophy-Determining Gene. *Cell* **1995**, *80*, 155–165.

23. Magenis, R. E.; Maslen, C. L.; Smith, L.; Allen, L.; Sakai, L. Y. Localization of the Fibrillin (FBN) Gene to Chromosome 15, Band q21.1. *Genomics* **1991**, *4*, 346–351.

24. Brenner, S.; Johnson, M.; Bridgham, J.; Golda, G.; Lloyd, D. H.; Johnson, D.; Luo, S.; McCurdy, S.; Foy, M.; Ewan, M.; Roth, R.; George, D.; Eletr, S.; Albrecht, G.; Vermaas, E.; Williams, S. R.; Moon, K.; Burcham, T.; Pallas, M.; DuBridge, R. B.; Kirchner, J.; Fearon, K.; Mao, J.; Corcoran, K. Gene Expression Analysis by Massively Parallel Signature Sequencing (MPSS) on Microbead Arrays. *Nat. Biotech.* **2000**, *18*, 630–634.

25. Moresco, E. M. Y.; Li, X.; Beutler, B. Going Forward With Genetics: Recent Technological Advances and Forward Genetics in Mice. *Am. J. Pathol.* **2013**, *182*, 1462–1473.

26. Ioannidis, J. P. A. Expectations, Validity, and Reality in Omics. *J. Clin. Epidemiol.* **2010**, *63*, 945–949.

27. Wang, D. G.; Fan, J. B.; Siao, C. J.; Berno, A.; Young, P.; Sapolsky, R.; Ghandour, G.; Perkins, N.; Winchester, E.; Spencer, J.; Kruglyak, L.; Stein, L.; Hsie, L.; Topaloglou, T.; Hubbell, E.; Robinson, E.; Mittmann, M.; Morris, M. S.; Shen, N.; Kilburn, D.; Rioux, J.; Nusbaum, C.; Rozen, S.; Hudson, T. J.; Lipshutz, R.; Chee, M.; Lander, E. S. Large-Scale Identification, Mapping, and Genotyping of Single-Nucleotide Polymorphisms in the Human Genome. *Science* **1998**, *280*, 1077–1082.

28. Pujol, A.; Mosca, R.; Farres, J.; Aloy, P. Unveiling the Role of Network and Systems Biology in Drug Discovery. *Trends Pharm. Sci.* **2010**, *31*(3), 115–123.

29. Ureta-Vidal, A.; Ettwiller, L.; Birney, E. Comparative Genomics: Genome-Wide Analysis in Metazoan Eukaryotes. *Nat. Rev. Genet.* **2003**, *4*, 251–262.

30. Heynen-Genel, S.; Pache, L.; Chanda, S. K.; Rosen, J. Functional Genomic and High-Content Screening for Target Discovery and Deconvolution. *Exp. Opin. Drug Discov.* **2012**, *7*, 955–968.

31. Zhu, J.; Adli, M.; Zou, J. Y.; Verstappen, G.; Coyne, M.; Zhang, X.; Durham, T.; Miri, M.; Deshpande, V.; De Jager, P. L.; Bennett, D. A.; Houmard, J. A.; Muoio, D. M.; Onder, T. T.;

Camahort, R.; Cowan, C. A.; Meissner, A.; Epstein, C. B.; Shoresh, N.; Bernstein, B. E. Genome-Wide Chromatin State Transitions Associated With Developmental and Environmental Cues. *Cell* **2013**, *152*, 642–654.

32. Schwanhäusser, B.; Busse, D.; Li, N.; Dittman, G.; Schuchhardt, J.; Wolf, J.; Chen, W.; Selbach, M. Global Quantification of Mammalian Gene Expression Control. *Nature* **2011**, *473*, 337–342.

33. Thompson, J. D., Schaeffer-Reiss, C., Ueffing, M., Eds. Functional Proteomics; Springer-Verlag: Berlin, 2008; ISBN: 978-1-59745-398-1.

34. Alberghina, L., Westerhoff, H. V., Eds. Topics in Current Genetics, Vol. 13; Systems Biology: Definitions and Perspectives; Springer-Verlag: Berlin, 2005.

35. Cowper-Sal, R.; Cole, M. D.; Karagas, M. R.; Lupien, M.; Moore, J. H. Layers of Epistasis: Genome-Wide Regulatory Networks and Network Approaches to Genome-Wide Association Studies. *Wiley Interdiscip. Rev. Syst. Biol. Med.* **2011**, *3*, 513–526.

36. Hartwell, L. H.; Hopfield, J. J.; Leibler, S.; Murray, A. W. From Molecular to Modular Cell Biology. *Nature* **1999**, *402*, C47–C52.

37. Penrod, N. M.; Cowper-Sal-lari, R.; Moore, J. H. Systems Genetics for Drug Target Discovery. *Trends Pharm. Sci.* **2011**, *32*, 623–630.

38. Dixon, S.; Stockwell, B. Identifying Druggable Disease-Modifying Gene Products. *Curr. Opin. Chem. Biol.* **2009**, *13*, 549–555.

39. Yildirim, M.; Goh, K. I.; Cusick, M. E.; Barabási, A. L.; Vidal, M. Drug-Target Network. *Nat. Biotechnol.* **2007**, *25*, 1119–1126.

40. Kim, J.; Gao, L.; Tan, K. Multi-Analyte Network Markers for Tumor Prognosis. *PLoS One* **2012**, *7*, e52973.

41. Gupta, G. P.; Nguyen, D. X.; Chiang, A. C.; Bos, P. D.; Kim, J. Y.; Nadal, C.; Gomis, R. R.; Manova-Todorova, K.; Massagué, J. Mediators of Vascular Remodelling Co-opted for Sequential Steps in Lung Metastasis. *Nature* **2007**, *446*, 765–770.

42. McDermott, J. E.; Diamond, D. L.; Corley, C.; Rasmussen, A. L.; Katze, M. G.; Waters, K. M. Topological Analysis of Protein Co-Abundance Networks Identifies Novel Host Targets Important for HCV Infection and Pathogenesis. *BMC Syst. Biol.* **2012**, *6*, 28.

43. Li, S.; Wu, L.; Zhang, Z. Constructing Biological Networks Through Combined Literature Mining and Microarray Analysis: A LMMA Approach. *Bioinformatics* **2006**, *22*, 2143–2150.

44. Battistella, M.; Mardsen, P. A. Advances, Nuances, and Potential Pitfalls When Exploiting the Therapeutic Potential of RNA Interference. *Clin. Pharm. Ther.* **2015**, *97*, 79–87.

45. Jinek, M.; Doudna, J. A. A Three-Dimensional View of the Molecular Machinery of RNA Interference. *Nature* **2009**, *457*, 405–412.

46. Ambros, V. The Functions of Animal MicroRNAs. *Nature* **2004**, *431*, 350–355.

47. Han, J.; Lee, Y.; Yeom, K. H.; Kim, Y. K.; Jin, H.; Kim, V. N. The Drosha-DGCR8 Complex in Primary DNA Processing. *Genes Dev.* **2004**, *18*, 3016–3027.

48. Foulkes, W. D.; Priest, J. R.; Duchaine, T. F. DICER1: Mutations, MicroRNAs and Mechanisms. *Nat. Rev. Cancer* **2014**, 662–672.

49. Hammond, S. M. An Overview of MicroRNAs. *Adv. Drug Deliv. Rev.* **2015**, XXX.

50. Kawamata, T.; Tomari, Y. Making RISC. *Trends Biochem. Sci.* **2010**, *35*, 368–376.

51. Swarts, D. C.; Makarova, K.; Wang, Y.; Nakanishi, K.; Ketting, R. F.; Koonin, E. V.; Patel, D. J.; van der Oost, J. The Evolutionary Journey of Argonaute Proteins. *Nat. Struct. Mol. Biol.* **2014**, *21*, 743–753.

52. Fabian, M. R.; Sonenberg, N. The Mechanics of miRNA-Mediated Gene Silencing: A Look Under the Hood of miRISC. *Nat. Struct. Mol. Biol.* **2012**, *19*, 586–593.

53. Meister, G.; Landthaler, M.; Peters, L.; Chen, P. Y.; Urlaub, H.; Luhrmann, R.; Tuschl, T. Identification of Novel Argonaute-Associated Proteins. *Curr. Biol.* **2005**, *15*, 2149–2155.

54. Wu, L.; Fan, J.; Belasco, J. G. MicroRNAs Direct Rapid Deadenylation of mRNA. *Proc. Natl. Acad. Sci. U. S. A.* **2006**, *103*, 4034–4039.

55. Ding, X. C.; Grosshans, H. Repression of *C. elegans* MicroRNA Targets at the Initiation Level of Translation Requires GW182 Proteins. *EMBO J.* **2009**, *28*, 213–222.

56. Hendrickson, D. G.; Hogan, D. J.; McCullough, H. L.; Myers, J. W.; Herschlag, D.; Ferrell, J. E.; Brown, P. O. Concordant Regulation of Translation and mRNA Abundance for Hundreds of Targets of a Human microRNA. *PLoS Biol.* **2009**, *7*, e1000238.

57. Fire, A.; Xu, S.; Montgomery, M.; Kostas, S.; Driver, S.; Mello, C. Potent and Specific Genetic Interference by Double-Stranded RNA in *Caenorhabditis elegans*. *Nature* **1998**, *391*, 806–811.

58. Elbashir, S. M.; Harborth, J.; Lendeckel, W.; Yalcin, A.; Weber, K.; Tuschl, T. Duplexes of 21-Nucleotide RNAs Mediate RNA Interference in Cultured Mammalian Cells. *Nature* **2001**, *411*, 494–498.

59. Ohkumo, T.; Masutani, C.; Eki, T.; Hanaoka, F. Use of RNAi in C. elegans. *Methods Mol. Biol.* **2008**, *442*, 129–137.

60. Flockhart, I.; Booker, M.; Kiger, A.; Boutros, M.; Armknecht, S.; Ramadan, N.; Richardson, K.; Xu, A.; Perrimon, N.; Mathey-Prevot, B. FlyRNAi: The Drosophila RNAi Screening Center Database. *Nucl. Acids Res.* **2006**, *34*, D489–D494.

61. Maillard, P. V.; Ciaudo, C.; Marchais, A.; Li, Y.; Jay, F.; Ding, S. W.; Voinnet, O. Antiviral RNA Interference in Mammalian Cells. *Science* **2013**, *342*, 235–238.

62. Sharp, P. A. RNAi and Double-Strand RNA. *Genes Dev.* **1999**, *13*, 139–141.

63. Chiu, Y. L.; Rana, T. M. siRNA Function in RNAi: A Chemical Modification Analysis. *RNA* **2003**, *9*, 1034–1048.

64. Rao, D. D.; Vorhies, J. S.; Senzer, N.; Nemunaitis, J. siRNA *vs.* shRNA: Similarities and Differences. *Adv. Drug Deliv. Rev.* **2009**, *61*, 746–759.

65. Rand, T. A.; Petersen, S.; Du, F.; Wang, X. Argonaute2 Cleaves the Anti-Guide Strand of siRNA During RISC Activation. *Cell* **2005**, *123*, 621–629.

66. Parker, J. S. How to Slice: Snapshots of Argonaute in Action. *Silence* **2010**, *1*, 3.

67. Clemens, J. C.; Worby, C. A.; Simonson-Leff, N.; Muda, M.; Maehama, T.; Hemmings, B. A.; Dixon, J. E. Use of Double-Stranded RNA Interference in *Drosophila* Cell Lines to Dissect Signal Transduction Pathways. *Proc. Natl. Acad. Sci. U. S. A.* **2000**, *97*, 6499–6503.

68. Draghici, B.; Ilies, M. A. Synthetic Nucleic Acid Delivery Systems: Present and Perspectives. *J. Med. Chem.* **2015**, *58*, 4091–4130.

69. Horn, T.; Sandmann, T.; Boutros, M. Design and Evaluation of Genome-Wide Libraries for RNA Interference Screens. *Genome Biol.* **2010**, *11*, R61.

70. Hu, K.; Lee, C.; Qiu, D.; Fotovati, A.; Davies, A.; Abu-Ali, S.; Wai, D.; Lawlor, E. R.; Triche, T. J.; Pallen, C. J.; Dunn, S. E. siRNA Library Screen of Human Kinases and Phosphatases Identifies Polo-Like Kinase 1 as a Promising New Target for the Treatment of Pediatric Rhabdomyosarcomas. *Mol. Cancer Ther.* **2009**, *8*, 3024–3035.

71. Laroche, G.; Giguere, P. M.; Roth, B. L.; Trejo, J. A.; Siderovski, D. P. RNA Interference Screen for RGS Protein Specificity at Muscarinic and Protease-Activated Receptors Reveals Bidirectional Modulation of Signaling. *Am. J. Physiol.* **2010**, *299*, C654–C664.

72. Arias Garcia, M.; Sanchez Alvarez, M.; Sailem, H.; Bousgouni, V.; Sero, J.; Bakal, C. Differential RNAi Screening Provides Insights Into the Rewiring of Signalling Networks During Oxidative Stress. *Mol. Biosyst.* **2012**, *8*, 2605–2613.

73. Griffiths, S. J.; Koegl, M.; Boutell, C.; Zenner, H. L.; Crump, C. M.; Pica, F.; Gonzalez, O.; Friedel, C. C.; Barry, G.; Martin, K.; Craigon, M. H.; Chen, R.; Kaza, L. N.; Fossum, E.;

Fazakerley, J. K.; Efstathiou, S.; Volpi, A.; Zimmer, R.; Ghazal, P.; Haas, J. A Systematic Analysis of Host Factors Reveals a Med23-Interferon-λ Regulatory Axis Against Herpes Simplex Virus Type 1 Replication. *PLoS Pathog.* **2013**, *9*, e1003514.

74. Jackson, A. L.; Linsley, P. S. Recognizing and Avoiding siRNA Offtarget Effects for Target Identification and Therapeutic Application. *Nat. Rev. Drug Discov.* **2010**, *9*, 57–67.

75. Ravi, D.; Wiles, A. M.; Bhavani, S.; Ruan, J.; Leder, P.; Bishop, A. J. A Network of Conserved Damage Survival Pathways Revealed by a Genomic RNAi Screen. *PLoS Genet.* **2009**, *5*, e1000527.

76. Friedman, A.; Tucker, G.; Singh, R.; Yan, D.; Vinayagam, A.; Hu, Y.; Binari, R.; Hong, P.; Sun, X.; Porto, M.; Pacifico, S.; Murali, T.; Finley, R.; Asara, J. M.; Berger, B.; Perrimon, N. Proteomic and Functional Genomic Landscape of Receptor Tyrosine Kinase and Ras to Extracellular Signal-Regulated Kinase Signaling. *Sci. Signal.* **2011**, *4*, rs10.

77. Wendler, F.; Gillingham, A. K.; Sinka, R.; Rosa-Ferreira, C.; Gordon, D. E.; Franch-Marro, X.; Peden, A. A.; Vincent, J. P.; Munro, S. A Genome-Wide RNA Interference Screen Identifies Two Novel Components of the Metazoan Secretory Pathway. *EMBO J.* **2010**, *29*, 304–314.

78. Neumüller, R. A.; Gross, T.; Samsonova, A. A.; Vinayagam, A.; Buckner, M.; Founk, K.; Hu, Y.; Sharifpoor, S.; Rosebrock, A. P.; Andrews, B.; Winston, F.; Perrimon, N. Conserved Regulators of Nucleolar Size Revealed by Global Phenotypic Analyses. *Sci. Signal.* **2013**, *6*, ra70.

79. Erhardt, S.; Mellone, B. G.; Betts, C. M.; Zhang, W.; Karpen, G. H.; Straight, A. F. Genome-Wide Analysis Reveals a Cell Cycle-Dependent Mechanism Controlling Centromere Propagation. *J. Cell Biol.* **2008**, *183*, 805–818.

80. Zhang, S.; Binari, R.; Zhou, R.; Perrimon, N. A Genomewide RNA Interference Screen for Modifiers of Aggregates Formation by Mutant Huntingtin in Drosophila. *Genetics* **2010**, *184*, 1165–1179.

81. Le Cong, L.; Ran, F. A.; Cox, D.; Lin, S.; Barretto, R.; Habib, N.; Hsu, P. D.; Wu, X.; Jiang, W.; Marraffini, L. A.; Zhang, F. Multiplex Genome Engineering Using CRISPR/Cas Systems. *Science* **2013**, *339*, 819–823.

82. Mali, P.; Yang, L.; Esvelt, K. M.; Aach, J.; Guell, M.; DiCarlo, J. E.; Norville, J. E.; Church, G. M. RNA-Guided Human Genome Engineering via Cas9. *Science* **2013**, *339*, 823–826.

83. Moore, J. D. The Impact of CRISPR–Cas9 on Target Identification and Validation. *Drug Discov. Today* **2015**, *20*, 450–457.

84. Ran, F. A.; Hsu, P. D.; Lin, C.-Y.; Gootenberg, J. S.; Konermann, S.; Trevino, A. E.; Scott, D. A.; Inoue, A.; Matoba, S.; Zhang, Y.; Zhang, F. Double Nicking by RNA-Guided CRISPR Cas9 for Enhanced Genome Editing Specificity. *Cell* **2013**, *154*, 1380–1389.

85. Cong, L.; Ann Ran, F.; Cox, D.; Lin, S.; Barretto, R.; Habib, N.; Hsu, P.D.; Wu, X.; Jiang, W.; Marraffini, L.A.; Zhang, F. Multiple Genome Engineering Using CRISPR/Cas Systems. Science, 2014, 339, 819–823.

86. Fu, Y.; Sander, J. D.; Reyon, D.; Cascio, V. M.; Joung, J. K. Improving CRISPR–Cas Nuclease Specificity Using Truncated Guide RNAs. *Nat. Biotechnol.* **2014**, *32*, 279–284.

87. Tsai, S. Q.; Wyvekens, N.; Khayter, C.; Foden, J. A.; Thapar, V.; Reyon, D.; Goodwin, M. J.; Aryee, M. J.; Joung, J. K. Dimeric CRISPR RNA-Guided FokI Nucleases for Highly Specific Genome Editing. *Nat. Biotechnol.* **2014**, *32*, 569–576.

88. Shalem, O.; Sanjana, N. E.; Zhang, F. Genome-Scale CRISPR–Cas9 Knockout Screening in Human Cells. *Science* **2014**, *343*, 84–87.

89. Zhou, Y.; Zhu, S.; Cai, C.; Yuan, P.; Li, C.; Huang, Y.; Wei, W. High-Throughput Screening of a CRISPR/Cas9 Library for Functional Genomics in Human Cells. *Nature* **2014**, *509*, 487–491.

90. Taylor, J.; Woodcock, S. A Perspective on the Future of High-Throughput RNAi Screening. Will CRISPR Cut Out the Competition or Can RNAi Help Guide the Way? *J. Biomol. Screen.* **2015**, *20*, 1040–1051.

91. Cong, F.; Cheung, A. K.; Huang, S.-M.A. Chemical Genetics-Based Target Identification in Drug Discovery. *Annu. Rev. Pharmacol. Toxicol.* **2012**, *52*, 57–78.

92. Vane, J. R. Inhibition of Prostaglandin Synthesis as a Mechanism of Action for Aspirin-Like Drugs. *Nature* **1971**, *231*, 232–235.

93. http://www.fda.gov/downloads/Drugs/GuidanceComplianceRegulatoryInformation/Guidances/ucm071597.pdf.

94. Munos, B. A Forensic Analysis of Drug Targets From 2000 Through 2012. *Clin. Pharmacol. Ther.* **2013**, *94*, 407–411.

95. Eder, J.; Sedrani, R.; Wiesmann, C. The Discovery of First-in-Class Drugs: Origins and Evolution. *Nat. Rev. Drug Discov.* **2014**, *13*, 577–587.

96. Ohlstein, E. H.; Ruffolo, R. R., Jr.; Elliott, J. D. Drug Discovery in the Next Millennium. *Annu. Rev. Pharmacol. Toxicol.* **2000**, *40*, 177–191.

97. Swinney, D. C.; Anthony, J. How Were New Medicines Discovered? *Nat. Rev. Drug Discov.* **2011**, *10*, 507–519.

98. Ramanan, V. K.; Saykin, A. J. Pathways to Neurodegeneration: Mechanistic Insights From GWAS in Alzheimer's Disease, Parkinson's Disease, and Related Disorders. *Am. J. Neurodegener. Dis.* **2013**, *2*, 145–175.

99. Seneci, P. Molecular Targets in Protein Folding and Neurodegenerative Disease; Academic Press: San Diego, 2014. 314 pp.

100. Seneci, P. Chemical Modulators in Protein Folding and Neurodegenerative Disease; Academic Press: San Diego, 2015. 260 pp.

101. Zhang, M.; Luo, G.; Zhou, Y.; Wang, S.; Zhong, Z. Phenotypic Screens Targeting Neurodegenerative Diseases. *J. Biomol. Screen.* **2014**, *19*, 1–16.

102. Ledgord, H. CRISPR, the Disruptor. *Nature* **2015**, *522*, 20–24.

103. Ralph, G. S.; Radcliffe, P. A.; Day, D. M.; Carthy, J. M.; Leroux, M. A.; Lee, D. C.; Wong, L. F.; Bilsland, L. G.; Greensmith, L.; Kingsman, S. M.; Mitrophanous, K. A.; Mazarakis, N. D.; Azzouz, M. Silencing Mutant SOD1 Using RNAi Protects Against Neurodegeneration and Extends Survival in an ALS Model. *Nat. Med.* **2005**, *11*, 429–433.

104. Yin, H.; Xue, W.; Chen, S.; Bogorad, R. L.; Benedetti, E.; Grompe, M.; Koteliansky, V.; Sharp, P. A.; Jacks, T.; Anderson, D. G. Genome Editing With Cas9 in Adult Mice Corrects a Disease Mutation and Phenotype. *Nat. Biotechnol.* **2014**, *32*, 551–553.

105. Van der Ree, M. H.; van der Meer, A. J.; de Bruijne, J.; Maan, R.; van Vliet, A.; Welzel, T. M.; Zeuzem, S.; Lawitz, E. J.; Rodriguez-Torres, M.; Kupcova, V.; Wiercinska-Drapalo, A.; Hodges, M. R.; Janssen, H. L. A.; Reesink, H. W. Long-Term Safety and Efficacy of MicroRNA Targeted Therapy in Chronic Hepatitis C Patients. *Antiviral Res.* **2014**, *111c*, 53–59.

106. Schenone, M.; Dančík, V.; Wagner, B. K.; Clemons, P. A. Target Identification and Mechanism of Action in Chemical Biology and Drug Discovery. *Nat. Chem. Biol.* **2013**, *9*, 232–240.

107. O'Connor, C. J.; Laraia, L.; Spring, D. R. Chemical Genetics. *Chem. Soc. Rev.* **2011**, *40*, 4332–4345.

108. Schreiber, S. L. Chemical Genetics Resulting From a Passion for Synthetic Organic Chemistry. *Bioorg. Med. Chem.* **1998**, *6*, 1127–1152.

109. Morgan, D. O. The Cell Cycle: Principles of Control, 2nd ed.; Oxford University Press: London, 2012.

110. Pardee, A. G1 Events and Regulation of Cell Proliferation. *Science* **1989**, *246*, 603–608.

111. Zetterberg, A.; Larsson, O.; Wiman, K. G. What Is the Restriction Point? *Curr. Opin. Cell Biol.* **1995**, *7*, 835–842.

112. Bell, S. P.; Dutta, A. DNA Replication in Eukaryotic Cells. *Annu. Rev. Biochem.* **2002**, *71*, 333–374.

113. Rieder, C. L.; Faruki, S.; Khodjakov, A. The Centrosome in Vertebrates: More Than a Microtubule-Organizing Center. *Trends Cell Biol.* **2001**, *11*, 413–419.

114. Zhang, N.; Kuznetsov, S. G.; Sharan, S. K.; Li, K.; Rao, P. H.; Pati, D. A Handcuff Model for the Cohesin Complex. *J. Cell Biol.* **2008**, *183*, 1019–1031.

115. Moseley, J. B.; Mayeux, A.; Paoletti, A.; Nurse, P. A Spatial Gradient Coordinates Cell Size and Mitotic Entry in Fission Yeast. *Nature* **2009**, *459*, 857–860.

116. Taylor, W. R.; Stark, G. R. Regulation of the G2/M Transition by p 53. *Oncogene* **2001**, *20*, 1803–1815.

117. De Souza, C. P.; Osmani, S. A. Mitosis, Not Just Open or Closed. *Eukaryot. Cell* **2007**, *6*, 1521–1527.

118. Pluta, A.; Mackay, A. M.; Ainsztein, A. M.; Goldberg, I. G.; Earnshaw, W. C. The Centromere: Hub of Chromosomal Activities. *Science* **1995**, *270*, 1591–1594.

119. Olson, M. O. J. The Nucleolus. Volume 15 of Protein Reviews; Springer: Berlin, 2011.

120. Kadauke, S.; Blobel, G. Mitotic Bookmarking by Transcription Factors. *Epigenetics Chromatin* **2013**, *6*, 6.

121. Bornens, M.; Azimzadeh, J. Origin and Evolution of the Centrosome. Eukaryotic Membranes and Cytoskeleton, *Adv. Exp. Med. Biol.* **2007**, *607*, 119–129.

122. Heald, R.; McKeon, F. Mutations of Phosphorylation Sites in Lamin A That Prevent Nuclear Lamina Disassembly in Mitosis. *Cell* **1990**, *61*, 579–589.

123. Chan, G.; Liu, S.; Yen, T. Kinetochore Structure and Function. *Trends Cell Biol.* **2005**, *15*, 589–598.

124. Cheeseman, I. M.; Desai, A. Molecular Architecture of the Kinetochore–Microtubule Interface. *Nat. Rev. Mol. Cell Biol.* **2008**, *9*, 33–46.

125. Lodish, H.; Berk, A.; Kaiser, C. A.; Krieger, M.; Bretscher, A.; Ploegh, H.; Amon, A.; Scott, M. P. Molecular Cell Biology, 7th ed.; MacMillan Higher Education: London, 2012.

126. Walczak, C. E.; Heald, R. Mechanisms of Mitotic Spindle Assembly and Function. *Int. Rev. Cytol.* **2008**, *265*, 111–158.

127. Mora Bermudez, F.; Matsuzaki, F.; Huttner, W. B. Specific Polar Subpopulations of Astral Microtubules Control Spindle Orientation and Symmetric Neural Stem Cell Division. *Elife* **2014**, *3*, e02875.

128. Tomomi, K. Mechanisms of Daughter Cell-Size Control During Cell Division. *Trends Cell Biol.* **2015**, *25*, 286–295.

129. Guo, Y.; Kim, C.; Mao, Y. New Insights Into the Mechanism for Chromosome Alignment in Metaphase. *Int. Rev. Cell Mol. Biol.* **2013**, *303*, 237–262.

130. Maiato, H.; DeLuca, J.; Salmon, E.; Earnshaw, W. The Dynamic Kinetochore-Microtubule Interface. *J. Cell Sci.* **2004**, *117*, 5461–5477.

131. Chan, G.; Yen, T. The Mitotic Checkpoint: A Signaling Pathway That Allows a Single Unattached Kinetochore to Inhibit Mitotic Exit. *Progr. Cell Cycle Res.* **2003**, *5*, 431–439.

132. Peters, J. M.; Tedeschi, A.; Schmitz, J. The Cohesin Complex and Its Roles in Chromosome Biology. *Genes Dev.* **2008**, *22*, 3089–3114.

133. Desai, A.; Maddox, P. S.; Mitchison, T. J.; Salmon, E. D. Anaphase A Chromosome Movement and Poleward Spindle Microtubule Flux Occur at Similar Rates in *Xenopus* Extract Spindles. *J. Cell Biol.* **1998**, *141*, 703–713.

134. Brust-Mascher, I.; Scholey, J. M. Mitotic Motors and Chromosome Segregation: The Mechanism of Anaphase B. *Biochem. Soc. Trans.* **2011**, *39*, 1149–1153.

135. Goldman, R. D.; Gruenbaum, Y.; Moir, R. D.; Shumaker, D. K.; Spann, T. P. Nuclear Lamins: Building Blocks of Nuclear Architecture. *Genes Dev.* **2002**, *16*, 533–547.

136. Hernandez-Verdun, D. The Nucleolus: Functional Organization and Assembly. *J. Appl. Biomed.* **2004**, *2*, 57–69.

137. Glotzer, M. The Molecular Requirements for Cytokinesis. *Science* **2005**, *307*, 1735–1739.

138. Albertson, R.; Riggs, B.; Sullivan, W. Membrane Traffic: A Driving Force in Cytokinesis. *Trends Cell Biol.* **2005**, *15*, 92–101.

139. Jordan, M. Mechanism of Action of Antitumor Drugs That Interact With Microtubules and Tubulin. *Curr. Med. Chem. Anti-Cancer Agents* **2012**, *2*, 1–17.

140. Molad, Y. Update on Colchicine and Its Mechanism of Action. *Curr. Rheumatol. Rep.* **2002**, *4*, 252–256.

141. Roussi, F.; Gueritte, F.; Fahy, J. The Vinca Alkaloids. In Anticancer Agents From Natural Products; Cragg, G. M., Kingston, D. G. I., Newman, D. J., Eds, 2nd ed.; Boca Raton: CRC Press, 2012; pp 177–198.

142. Abal, M.; Andreu, J. M.; Barasoain, I. Taxanes: Microtubule and Centrosome Targets, and Cell Cycle Dependent Mechanisms of Action. *Curr. Cancer Drug Targets* **2003**, *3*, 193–203.

143. Taylor, S.; Peters, J.-M. Polo and Aurora Kinases—Lessons Derived From Chemical Biology. *Curr. Opin. Cell Biol.* **2008**, *20*, 77–84.

144. Rath, O.; Kozielski, F. Kinesins and Cancer. *Nat. Rev. Cancer* **2012**, *12*, 527–539.

145. http://www.pfizer.com/news/press-release/press-release-detail/pfizer_receives_u_s_fda_accelerated_approval_of_ibrance_palbociclib.

146. Kollareddy, M.; Zheleva, D.; Dzubak, P.; Brahmkshatriya, P. S.; Lepsik, M.; Hajduch, M. Aurora Kinase Inhibitors: Progress Towards the Clinic. *Invest. New Drugs* **2012**, *30*, 2411–2432.

147. Yim, H. Current Clinical Trials With Polo-Like Kinase 1 Inhibitors in Solid Tumors. *Anticancer Drugs* **2013**, *24*, 999–1006.

148. Sánchez-Martínez, C.; Gelbert, L. M.; Lallena, M. J.; de Dios, A. Cyclin Dependent Kinase (CDK) Inhibitors as Anticancer Drugs. *Bioorg. Med. Chem. Lett.* **2015**, https://doi.org/10.1016/j.bmcl.2015.05.100.

149. Stanton, R. A.; Gernert, K. M.; Nettles, J. H.; Aneja, R. Drugs That Target Dynamic Microtubules: A New Molecular Perspective. *Med. Res. Rev.* **2011**, *31*, 443–481.

150. Mayer, T. U.; Kapoor, T. M.; Haggarty, S. J.; King, R. W.; Schreiber, S. L.; Mitchison, T. J. Small Molecule Inhibitor of Mitotic Spindle Bipolarity Identified in a Phenotype-Based Screen. *Science* **1999**, *286*, 971–974.

151. Haggarty, S. J.; Mayer, T. U.; Miyamoto, D. T.; Fathi, R.; King, R. W.; Mitchison, T. J.; Schreiber, S. L. Dissecting Cellular Processes Using Small Molecules: Identification of Colchicine-Like, Taxol-Like and Other Small Molecules That Perturb Mitosis. *Chem. Biol.* **2000**, *7*, 275–286.

152. Vincent, I.; Rosado, M.; Davies, P. Mitotic Mechanisms in Alzheimer's Disease? *J. Cell Biol.* **1996**, *132*, 413–425.

153. Anderson, H. J.; de Jong, G.; Vincent, I.; Roberge, M. Flow Cytometry of Mitotic Cells. *Exp. Cell Res.* **1998**, *238*, 498–502.

154. Stockwell, B. R.; Haggarty, S. J.; Schreiber, S. L. High-Throughput Screening of Small Molecules in Miniaturized Mammalian Cell-Based Assays Involving Post-Translational Modifications. *Chem. Biol.* **1999**, *6*, 71–83.

155 DiverSet E, ChemBridge Corporation, http://www.chembridge.com/screening_libraries/diversity_libraries/.

156. Mitchison, T.; Kirschner, M. Microtubule Assembly Nucleated by Isolated Centrosomes. *Nature* **1984**, *312*, 232–237.

157. Desai, A.; Verma, S.; Mitchison, T. J.; Walczak, C. E. Kin1 Kinesins Are Microtubule-Destabilyzing Enzymes. *Cell* **1999**, *96*, 69–78.

158. Drewe, J.; Kasibhatla, S.; Tseng, B.; Shelton, E.; Sperandio, D.; Yee, R. M.; Litvak, J.; Sendzik, M.; Spencer, J. R.; Cai, S. X. Discovery of 5-(4-Hydroxy-6-methyl-2-oxo-2H-pyran-3-yl)-7-phenyl-(E)-2,3,6,7-tetrahydro-1,4-thiazepines as a New Series of Apoptosis Inducers Using a Cell- and Caspase-Based HTS Assay. *Bioorg. Med. Chem.* **2007**, *17*, 4987–4990.

159. Cramer, L. P.; Mitchison, T. J.; Theriot, J. A. Actin-Dependent Motile Forces and Cell Motility. *Curr. Opin. Cell Biol.* **1994**, *6*, 82–86.

160. Rieder, C. L.; Davison, E. A.; Jensen, L. C.; Cassimeris, L.; Salmon, E. D. Oscillatory Movements of Monooriented Chromosomes and Their Position Relative to the Spindle Pole Result From the Ejection Properties of the Aster and Half-Spindle. *J. Cell Biol.* **1986**, *103*, 581–591.

161. Heck, M. M. S.; Pereira, A.; Pesavento, P.; Yannoni, Y.; Spradling, A. C. The Kinesin-Like Protein KLPGIF Is Essential for Mitosis in Drosophila. *J. Cell Biol.* **1993**, *723*, 665–679.

162. Enos, A. P.; Morris, N. R. Mutation of a Gene That Encodes a Kinesin-Like Protein Blocks Cell Division in A. *nidulans. Cell* **1990**, *60*, 1019–1027.

163. Hagan, I.; Yanagida, M. Novel Potential Mitotic Motor Protein Encoded by the Fission Yeast cut7+ Gene. *Nature* **1990**, *347*, 563–566.

164. Sawin, K. E.; Leguellec, K.; Philippe, M.; Mitchison, T. J. Mitotic Spindle Organization by a Plus-End-Directed Microtubule Motor. *Nature* **1992**, *359*, 540–543.

165. Blangy, A.; Lane, H. A.; d'Herin, P.; Harper, M.; Kress, M.; Nigg, E. A. Phosphorylation by p34cdc2 Regulates Spindle Association of Human Eg5, a Kinesin-Related Motor Essential for Bipolar Spindle Formation *In Vivo. Cell* **1995**, *83*, 1159–1169.

166. Cole, D. G.; Saxton, W. M.; Sheehan, K. B.; Scholey, J. M. A "Slow" Homotetrameric Kinesin-Related Motor Protein Purified From Drosophila Embryos. *J. Biol. Chem.* **1994**, *269*, 22913–22916.

167. Kashina, A. S.; Baskin, R. J.; Cole, D. G.; Wedaman, K. P.; Saxton, W. M.; Scholey, J. M. A Bipolar Kinesin. *Nature* **1996**, *379*, 270–272.

168. Kapoor, T. M.; Mitchison, T. J. Allele-Specific Activators and Inhibitors for Kinesin. *Proc. Natl. Acad. Sci. U. S. A.* **1999**, *96*, 9106–9111.

169. Kapoor, T. M.; Mayer, T. U.; Coughlin, M. L.; Mitchison, T. J. Probing Spindle Assembly Mechanisms With Monastrol, a Small Molecule Inhibitor of the Mitotic Kinesin Eg5. *J. Cell Biol.* **2000**, *150*, 975–988.

170. Maliga, Z.; Kapoor, T. M.; Mitchison, T. J. Evidence That Monastrol Is an Allosteric Inhibitor of the Mitotic Kinesin Eg5. *Chem. Biol.* **2002**, *9*, 989–996.

171. Cochran, J. C.; Gatial, J. E., III; Kapoor, T. M.; Gilbert, S. P. Monastrol Inhibition of the Mitotic Kinesin Eg5. *J. Biol. Chem.* **2005**, *280*, 12658–12667.

172. Yan, Y.; Sardana, V.; Xu, B.; Homnick, C.; Halczenko, W.; Buser, C. A.; Schaber, M.; Hartman, G. D.; Huber, H. E.; Kuo, L. C. Inhibition of a Mitotic Motor Protein: Where, How, and Conformational Consequences. *J. Mol. Biol.* **2004**, *335*, 547–554.

173. DeBonis, S.; Simorre, J.-P.; Crevel, I.; Lebeau, L.; Skoufias, D.; Blangy, A.; Ebel, C.; Gans, P.; Cross, R.; Hackney, D. D.; Wade, R. H.; Kozielski, F. Interaction of the Mitotic Inhibitor Monastrol With Human Kinesin. *Biochemistry* **2003**, *42*, 338–349.

174. Sarli, V.; Giannis, A. Targeting the Kinesin Spindle Protein: Basic Principles and Clinical Implications. *Clin. Cancer Res.* **2008**, *14*, 7853–7857.

175. Chu, Q. S.; Holen, K. D.; Rowinsky, E. K.; Alberti, D. B.; Monroe, P.; Volkman, J. L.; Hodge, J. P.; Sabry, J.; Ho, P. T. C.; Wilding, G. A Phase I Study to Determine the Safety and Pharmacokinetics of IV Administered SB-715992, a Novel Kinesin Spindle Protein (KSP) Inhibitor in Patients With Solid Tumors. *Proc. Am. Soc. Clin. Oncol.* **2003**, *22*, 525.

Chapter 2

Step II: Target Validation

Molecular targets are the foundation of pharmaceutical research, but an R&D project requires *target validation* (TV, phase 2, Fig. 2.1) to reinforce the disease-target connection, and to reduce the risk of failure in later phases.

The definition of a validated target is straightforward: at least one marketed drug exerts its therapeutic action through its modulation.[1] Pursuing a validated molecular target with novel, patentable chemical classes (me-too drugs[2]) minimizes failure due to lack of efficacy, or to target-related toxicity effects. Unfortunately, many diseases show a high unmet medical need.[3] Effective drugs against them are unavailable, mostly due to the lack of relevant, validated molecular targets.

Working on validated targets entails lower risk of failure, lower costs and timelines, identification of second- and third-generation drugs with optimized safety and efficacy profiles. Innovative mechanisms of action address unmet medical needs and have huge upside value when leading to clinical success. Pharmaceutical companies privilege validated targets to minimize the risk and replenish their clinical pipelines. Public funding, academic research, and biotech-driven innovation should tackle innovative targets and contribute to reduce their risk.[4]

The average success rate of a pharmaceutical R&D project, defined as the percentage of projects that produce a drug approved for market launch, is estimated at an abysmal $\approx 4\%$.[5] Difficult therapeutic areas show an even lower success rate ($\approx 0.5\%$, Alzheimer's Disease[6]). Most failures happen during clinical development ($\approx 90\%$), in particular in Phase II/efficacy studies ($\approx 68\%$).[7] Around 35% of R&D projects against a molecular target are stopped before entering clinical development.[5] These numbers seem to indicate that preclinical research is much more efficient and successful than clinical research, but that is not true.

Phase II/Phase III clinical failure is mostly due to lack of efficacy in patients ($\approx 56\%$), and to safety/toxicity issues ($\approx 28\%$).[8] Failing clinical candidates show in vivo target occupancy/engagement, a good pharmacokinetic (PK)/pharmacodynamic (PD) profile, lack of toxicity and solid efficacy in animal models.[1] Their failure is often reconducible to incomplete TV in early discovery (i.e., limited assessment of human epidemiologic data) and preclinical research (i.e., animal models that do not recapitulate the human disease). Increased efforts in early TV should decrease clinical attrition rates and increase resource/budget

Chemical Sciences in Early Drug Discovery. https://doi.org/10.1016/B978-0-08-099420-8.00002-X

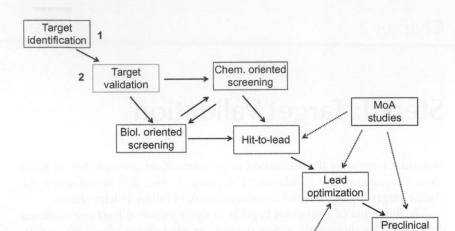

FIG. 2.1 The R&D pharmaceutical process.

savings. Budget savings should be allocated to basic research, increasing the number of identified and validated targets as a gateway to first-in-class drugs against diseases with high unmet medical need.[5]

2.1. THE FOUNDATION: MOLECULAR BIOLOGY, ONCE MORE NO CHEMISTRY

The identification of a molecular target, using tools and techniques described in Chapter 1, partially overlaps with its validation. A robust target-disease connection must be established to justify a high-throughput screening (HTS) campaign, or a long rational drug design approach (phases 3 and 4, Fig. 2.1). The duration of TV is blurred, as the validation of a target-disease connection increases until drug approval, or is disproven when a development candidate fails. Even when published data seem robust, their reliability may be questionable[9–11] due to lack of reproducibility for in vitro and in vivo TV methods.

Physiological (i.e., corticosteroid-driven amelioration of rheumatoid arthritis in pregnant women[12]) and traditional-medicine-driven observations (i.e., inflammation-reducing effects of extracts from willow bark[13]) validated natural products (NPs) and extracts. Pathological correlations (i.e., increased cholesterol-driven risk of heart disease[14]) identified disease-connected pathways. In all cases, the associated Molecular targets were identified later, with no need for further validation.

Cloning identified disease-determining translocations, as the one regarding chromosomes 9 and 22 in chronic myelogenous leukemia (CML) patients.[15]

The resulting breakpoint cluster region-Abelson/BCR-ABL fusion onco-gene[16] showed higher tyrosine kinase activity, causing leukemia-like disease in mice.[17] Massive parallel sequencing established genome-wide association studies (GWAS)[18] and genome-wide regulatory networks (GWRN).[19] The -omics revolution led to gene expression profiles (transcriptomics), protein expression profiles (proteomics), reversible DNA and histone modification profiles (epigenomics), transcription-factor-binding profiles (cistromics), and small-molecule (SM)/metabolite profiles (metabolomics).[20] Systems biology built complex networks and extracted information to identify, and sometimes to validate targets in specific subpopulations of patients.[20]

Biology-driven validation methods take place early in an R&D project, but may extend up to preclinical studies. *Human epidemiological data* are relevant to clinical use, but do not discriminate between causative genes/targets (relevant to the disease, i.e., HMG-CoA reductase inhibition[21] and low-density lipoprotein-cholesterol decrease[22] against heart disease) and noncausative genes/targets (consequence of the disease, i.e., cholesteryl ester transfer protein inhibition[23] and high-density lipoprotein increase[24] against heart disease). Their relevance is checked by generating *models* that describe the phenotype resulting from modulation of a gene in *cells* and in *living organisms*. Cellular TV is simple, should be done in early R&D, but has limited human relevance.[1] *Drosophila*,[25] zebrafish,[26] and rodent models[27] have higher human relevance,[28] but are used later in the R&D process.

Irreversible TV models are induced in vitro or in vivo by a loss-of-function (LOF) gene knockout,[29] or by a gain-of-function (GOF) gene knockin.[30] Recently, *CRISPR-Cas9/gene editing*-induced TV models have been reported.[31] Reversible TV models are induced by protein/antibody knockdown (LOF up to a few years[32]), and by *RNA interference* (RNAi, LOF up to a few weeks[33]). RNAi and CRISPR-Cas9/gene editing were technically described in Chapter 1.

2.2. CHEMISTRY IN MODERN TARGET VALIDATION: PHENOTYPIC SCREENING, REVERSE CHEMICAL GENETICS

The true validation of a molecular target is achieved in clinical trials, either using biomarkers[34] or drug candidates. Earlier, intermediate validation studies must be carried out to focus on prospective targets in areas with high unmet medical need.

Biology-driven target identification (TI) and TV methods rely on the connection between a diseased phenotype and a putative molecular target. The effects of an abnormal protein, of a mutated/edited/ablated gene in clinical isolates, in cellular or in animal models establish a direct target-disease connection. An indirect connection is established from a phenotype-impacting siRNA sequence and its target mRNA sequence.

Chemical (SMs) or biological modulators (proteins, protein domains, antibodies) of a gene product are the core of *chemistry-driven TV methods*.[35]

A cell-permeable, target-selective chemical or biological probe shows concentration-dependent and reversible modulation of a disease phenotype. It shows if target modulation is linked to any toxic effect in cellular or animal models, and discriminates between causative (curative effects in disease models) and noncausative targets (lack of curative effects in disease models).[35]

Chemistry-based TV includes endogenous molecules (e.g., therapeutic improvements observed during pregnancy for rheumatoid arthritis patients due to elevated cortisol levels) that became drugs well before the -omics revolution.[36] Even earlier, NPs were used as probes to discover targets and target pathways in human diseases.[37] Some synthetic drugs are even today approved for medical usage without a confirmed MoA[38] that may be later determined.[39] Novel targets can be identified by drug repurposing,[40] i.e., by searching additional curative effects in phenotypic screenings for approved drugs and clinical candidates. This Section describes the use of phenotypic modulators from forward chemical genetics (FCG) screening campaigns to validate novel molecular targets.

Once a phenotypic modulator/FCG hit is discovered, the elucidation of its MoA (i.e., a novel target-disease connections) is time- and labor intensive.[41]

An FCG hit is the gateway to *virtual*[42–44] and *tangible*[41,45,46] *target deconvolution methods* (Fig. 2.2).

Virtual target deconvolution methods use the hit structure and its biological effects to search accessible sources (scientific literature, bio- and chemoinformatic DBs and softwares, etc.) for chemical and biological similarity. Tangible efforts entail either chemical modification of the FCG hit (detection-friendly in vitro and/or in vivo interaction with the target), or its use in conjunction with biology-driven TV methods (gene expression and protein interaction profiles, RNAi, CRISPR-Cas9).

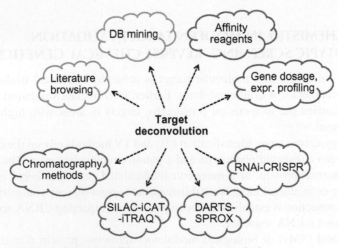

FIG. 2.2 Virtual *(top left)* and tangible target deconvolution methods.

Chemical similarity (see Chapter 3) is estimated by computational programs[47] that generate compound-specific fingerprints using molecular descriptors.[48,49] The Tanimoto index[50] represents the structural similarity of two compounds, varying between 0 (extremely diverse) and 1 (identical). Similar compounds could show similar biological effects, and share the same molecular targets,[51] but that is not always true.[52,53]

The comparison of an FCG hit with large (>1 M) sets of biologically active molecules in public DBs[43] could provide information on its putative target(s) if highly similar compounds (Tanimoto index ≥0.85) are found.[54] A similarity-based comparison of >3000 drugs with >65,000 ligands yielded several putative ligand-target associations.[55] 30 among the strongest associations were tested, 23 (>75%) were experimentally confirmed ($K_i \leq 15\,\mu M$), and 13 (>30%) resulted in submicromolar potency on the target.[55]

The effects of drug-like compounds on whole biological systems/networks, labeled by various groups as computational chemogenomics,[56] proteochemometric modeling,[57] network pharmacology[58] and systems chemical biology,[59] are stored in public and private molecular information systems.[44] They contain structural information on compounds, proteins, and nucleic acids together with protein-protein, protein-nucleic acid, and protein-ligand interaction data in many cell lines. Available semantic algorithms[60] facilitate the integrated browsing of system chemical biology DBs, modeling relationships between molecules, and their effects in biological environments. The ultimate goal is a virtual pharmacological space,[61] containing network-connected billions of drug-like SMs and thousands of putative molecular targets. Such space should be captured in a searchable DB suitable for multitargeted structure–activity relationship (SAR) determination. The wordle shown in Fig. 2.3[62] underlines the disorganized nature of such virtual pharmacological space.

The available information is growing, but the quality of stored data is sometimes questionable.[63] The definition and implementation of standards

FIG. 2.3   Virtual pharmacological space: components, methods, descriptors.

for uploaded data in public DBs such as ChEMBL,[64] PubChem,[65] or the Connectivity Map[66] is ongoing, to increase their usefulness.

Once an FCG hit is identified and characterized, its biological signature (impact on gene/protein expression and functionality, impact on post-translational modifications (PTMs) of proteins, interaction maps with endogenous molecules, cellular localization, etc.) is determined. Then, it is compared with the virtual pharmacological space, to determine similarities with known compounds, affinities for validated and putative targets, and for toxicity-inducing proteins. Virtual methods are less cost- and labor intensive than tangible methods, and often narrow the field of putative targets for an FCG hit (focus on a pathway, or an enzyme/receptor class).[44]

An FCG hit is then characterized using *tangible methods*,[67] if only to experimentally confirm (or disprove) computational predictions, by observing putative target-FCG hit interactions. Most frequently, targets are validated ex novo by testing an FCG hit on relevant models/cell lines/systems. Such profiling discriminates between causative/phenotype-affecting targets and noncausative/phenotype-irrelevant proteins binding to the FGC hit.

The effects of an FCG hit can be observed on a global/-omics perspective in cellular assays (mostly in yeasts), where its target proteins are dosed via genetic intervention (*gene dosage*).[45] *Haploinsufficiency profiling* (*HIP*[68]) entails the growth, in presence of the FCG hit, of a yeast cell library bearing bar-coded heterozygous deletions for each essential gene. Cells containing heterozygous gene deletions targeted by the FCG hit show impaired growth. They are identified after amplification by polymerase chain reaction (PCR), comparison with genomic DNA from untreated cells, and identification of low abundance/target genes (Fig. 2.4, *left*). Using HIP, the impairment of RNA processing was identified as the primary MoA of 5-fluorouracil.[69]

Homozygous deletion profiling (*HOP*[70]) adds an FCG hit to a bar-coded yeast cell library where nonessential genes are fully deleted. The FCG hit inhibits the growth of cells lacking a protein regulator of their essential protein target, resulting in higher sensitivity/growth inhibition. The protein regulator is identified as seen for HIP. The essential protein target is extracted through comparison of HOP profiling with phenotypes caused by genetic mutations (Fig. 2.4, *middle*). Using HOP, other cellular targets of microtubule-stabilizing peloruside A were identified, and its complex cellular activity was explained.[71] *Multicopy suppression profiling* (*MSP*[72]) uses FCG hit-resistant cells transformed with a bar-coded plasmid library of wild-type (WT) cell genes. In presence of an FCG hit, the cell subpopulation transformed with the WT target(s) becomes sensitive to the FCG hit, and its growth is inhibited. PCR amplification of bar-coded WT target genes, and comparison with genomic DNA from untreated cells led to their identification as low abundance/target genes (Fig. 2.4, *right*). Using MSP, the MoA of theopalauamide was clarified (binding to ergosterol, disruption of ergosterol-containing membranes).[73] Microarray-based HT technologies allowed the joint use of HIP, HOP, and MSP to identify the phenotype-connected molecular targets of 188 SMs.[74]

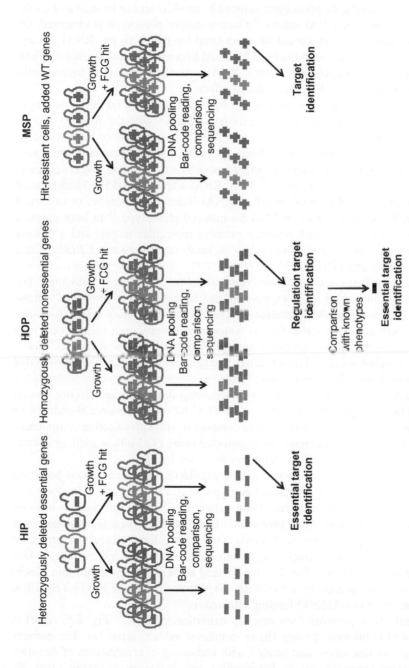

FIG. 2.4 Gene dosage in TV: HIP, HOP, and MSP.

RNAi can be coupled with FCG to determine the MoA of FCG hits. An RNAi-induced phenotype can be used as a phenotypic screening to find SM hits.[75] Conversely, the phenotype induced by an FCG hit can be used as a model in a genome-wide RNAi screen.[76] Once a similar phenotype is observed, the RNAi-impacted target should be modulated by the FCG hit. RNAi library-silenced cells treated with the FCG hit could become resistant, further confirming the FCG hit-target connection.[77] RNAi screens can be run in human cells, further validating the FCG hit-induced effect.[78]

The effects induced by an SM and by RNAi silencing on the same target/phenotype differ in terms of kinetics (much faster with the former) and efficacy (depending on binding strength for SMs, on partial or total silencing for RNAi). If the target is multifunctional (i.e., a scaffold protein with enzymatic activity), RNAi silencing fully inactivates it, while an SM specifically modulates one of its functions.[79] Thus, RNAi silencing and FCG can be used to screen for an SM modifier of an RNAi-induced phenotype, or to screen for an RNAi modifier of an FCG hit-induced phenotype.[25] In both cases, a connection is established between putative molecular targets and a disease phenotype. Similar connections can be established between *CRISPR-Cas9 gene editing* and FCG.[80]

FCG hits are used in *transcriptomic/mRNA profiling*,[81] providing hit-dependent changes in gene expression, and in *metabolomic profiling*,[82] showing hit-induced metabolic changes. FCG hit-driven profiling takes advantage of modern detection technologies. Automated fluorescence microscopy with multiple probes,[83] and impedance measurements with electrical sensors[84] determine morphological and quantitative changes, and clarify the effects of active compounds on cellular processes.

Proteomic profiling[85] uses two-dimensional difference gel electrophoresis (2D-DIGE) to quantify the effect of an FCG hit on the relative abundance of fluorescently tagged proteins, and to compare it with known active compounds. FCG hit-protein interactions can be detected using FCG hits as such, or chemically modified probes (chemical proteomics, see later).

Drug affinity response target stability (*DARTS*,[86] Fig. 2.5, *left*) is based on the increased stability of ligand-bound proteins. A cell lysate is incubated in presence (step 1b) or in absence (step 1a) of an FCG hit. Proteolysis is induced by treatment with multiple proteases[87] (step 2). The lysate is separated, and comparative quantitative proteomics (step 3) highlight proteolysis-protected putative targets. Their structure is deconvoluted using mass spectrometry (MS)-based methods (step 4, Fig. 2.5, *left*). Using DARTS, the F-box protein Met30 was identified as a target in the Met30-Skp1 complex for the FCG hit SMER3, excluding a Skp1-SMER3 binding interaction.[88]

Stability of proteins from rates of oxidation (*SPROX*,[89] Fig. 2.5, *right*) is run in FCG hit-treated (step 1b) or -untreated extracts (step 1a). The extracts are split in test tubes, and treated with increasing concentrations of denaturing agent (guanidinium hydrochloride) and hydrogen peroxide (step 2).

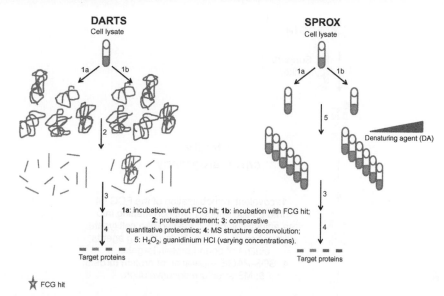

FIG. 2.5   FCG hit-induced effects on target stability: DARTS and SPROX.

Denaturation-dependent S-oxidation of Met residues takes place. FCG hit-bound proteins are more resistant to denaturation, and their oxidation requires higher concentrations of denaturing agents. Quantitative proteomics detect Met oxidation patterns in treated and untreated extracts (step 3). MS-based structural deconvolution (step 4, Fig. 2.5, *right*) identifies FCG-hit bound, Met-containing target proteins. SPROX works with abundant targets strongly bound to the FCG hit.[67] SPROX was used to identify novel cyclosporine A targets.[89]

Methods relying on FCG hits as such include *target identification by chromatographic coelution (TICC[90])* that takes advantage of varying liquid chromatography (LC) retention times for free and ligand-bound target proteins. *Size exclusion chromatography for target identification (SEC-TID[91])* compares the SEC profile of a library of purified proteins in presence or absence of an FCG hit, and identifies putative targets by changes in their apparent size.

The interaction between a modified FCG hit and its target(s) can be observed by *chemical proteomics*.[92] Chemical proteomics require the chemical conversion of a hit into a detectable probe, without losing target affinity. The FCG hit can be immobilized onto a support, or can be used as a soluble and cell-permeable probe after modification.

Affinity chromatography[93] entails the immobilization of an FCG hit onto a support through a linker (step 1, Fig. 2.6). The supported FCG hit is loaded in a column and used in the chromatographic separation of a biological extract (step 2). After elimination of unbound eluates, treatment with a solution of denaturing agents elutes the putative targets (step 3). They are separated by sodium dodecyl sulfate-polyacrylamide gel electrophoresis (SDS-PAGE,

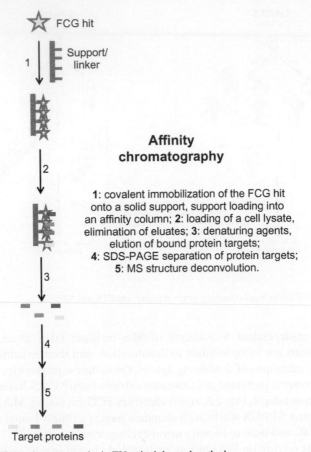

Affinity chromatography

1: covalent immobilization of the FCG hit onto a solid support, support loading into an affinity column; 2: loading of a cell lysate, elimination of eluates; 3: denaturing agents, elution of bound protein targets; 4: SDS-PAGE separation of protein targets; 5: MS structure deconvolution.

Target proteins

FIG. 2.6 Affinity chromatography in TV: principles and methods.

step 4), and structurally deconvoluted by MS-based methods (step 5, Fig. 2.6). Affinity chromatography is simple, but many factors influence the quality of its results.

The functionalization of the FCG hit must take place on a position that is not involved in target binding. The FCG hit **2.1** (Fig. 2.7) was selected in a phenotypic screening as a suppressor of cytokine-induced β-cell apoptosis from a diastereomeric benzoxazocinone library **2.2a-d**.[94,95]

A potent, nontoxic compound **2.3**[96] was identified after limited exploration of the urea and sulfonamide groups, and was selected for MoA/TV studies. The compound was coupled with an aminopolyethyleneglycol (aminoPEG)-based linker through its hydroxyl group.[80] The aminocarbamate **2.4** retained the biological activity of **2.3**, and its immobilization onto agarose beads yielded the affinity chromatography support **2.5** (Fig. 2.7). A kinase-independent inhibition of JAK-STAT signaling was determined as the MoA of **2.3** through **2.5**-driven affinity chromatography.[80]

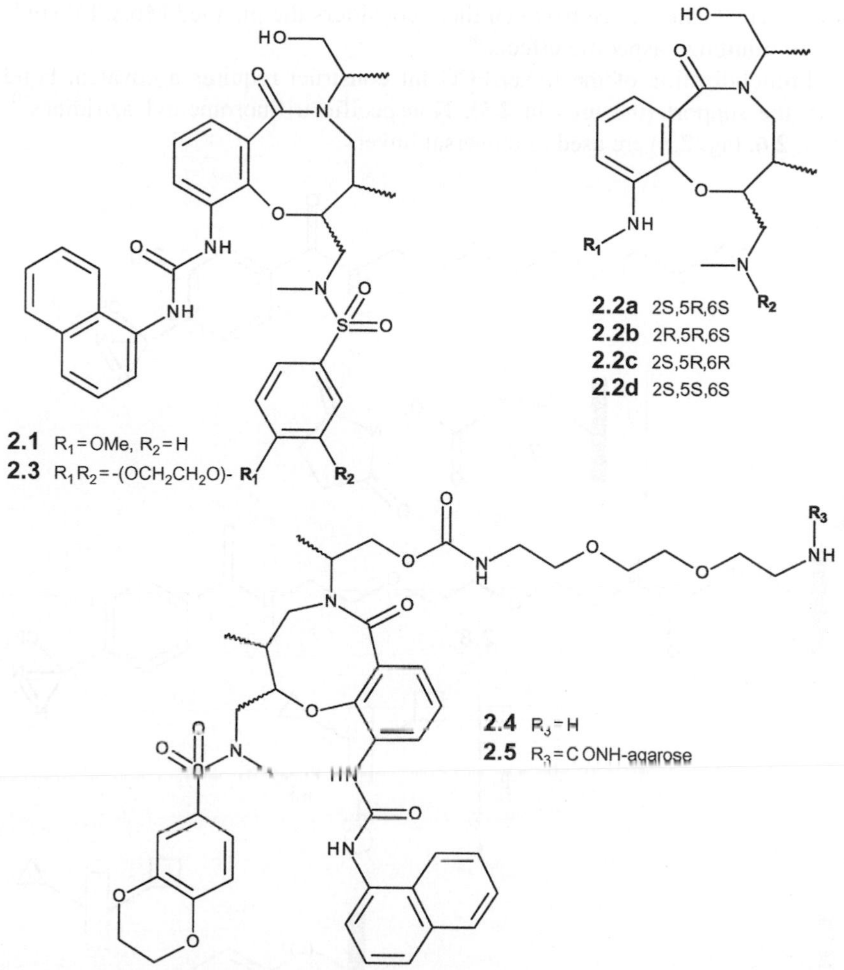

FIG. 2.7 Kinase-independent inhibition of JAK-STAT signaling: structure of FCG hits, chemical probes, and affinity reagents.

The linker and the solid support in an immobilized affinity compound must be biologically inert, to avoid aspecific protein fishing. The linker must be long enough to avoid steric hindrance between the target and the support.[93] It must be hydrophilic enough (e.g., PEG-based as in **2.4** or longer, tartaric acid-based[97]) to prevent autoaggregation of the immobilized affinity compound and nonspecific protein binding. Rigid portions as polyproline helices[98] are introduced for similar purposes. Solid supports include sugar-based agarose (as in **2.5**) and sepharose, polymethylacrylate-based resins,[99] and magnetic nanoparticles coated by polyglycidyl methacrylate.[100] They show varying hydro- and lipophilicity, stability in organic and aqueous media, and chemical compatibility with the immobilization reaction of a linker-FCG hit construct.[101] The choice between them considers the planned MoA/TV studies, to minimize aspecific effects.[46]

Immobilization of the linker-FCG hit construct requires a covalent bond with the support (the urea in **2.5**). Nonspecific trifluoromethyl aziridines[102] (e.g., **2.6**, Fig. 2.8) are used as universal linkers.

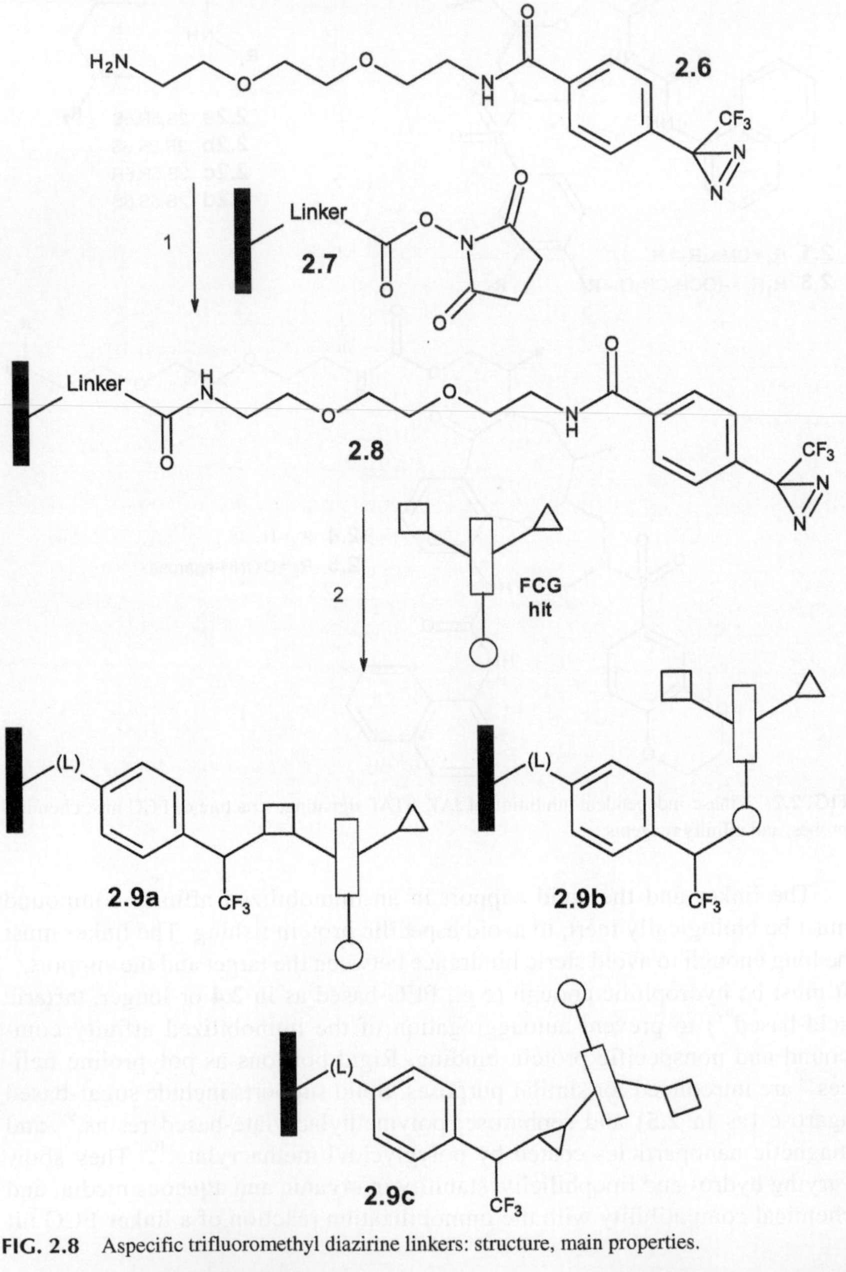

FIG. 2.8 Aspecific trifluoromethyl diazirine linkers: structure, main properties.

Trifluoromethyl aziridine linker **2.6** is stable, once covalently immobilized onto a solid support **2.7**[103] to give **2.8** (step 1). Upon UV irradiation, nitrogen loss produces highly reactive carbenes that react aspecifically in presence of a biologically active FCG hit (step 2), even when it does not contain reactive groups. Carbene insertion connects the linker to different regions of the FCG hit (e.g., **2.9a-c**, Fig. 2.8). Some linker-FCG hit connections will cause loss of biological activity, but others will retain it and will be used in affinity chromatography. Conventional supports[104] or microarrays[105] are compatible, with the latter enabling the use of photoaffinity linkers for HT chemical genetics studies.

The interaction between an immobilized affinity reagent and a target in a biological extract (≥ 1 M proteins) is influenced by many factors.[46] The best scenario entails a potent immobilized compound ($<1\text{-}\mu$M affinity) and an abundant target. An FCG hit, though, may interact with multiple cellular proteins; its phenotypic effects may be due to weak interactions with abundant proteins; or to strong interactions with low abundance targets. Moreover, an MoA/TV project may be plagued by aspecific interactions with ubiquitous proteins.

Target specificity can be attained by prefractionation of a cell lysate, if the cellular localization of targets for an FCG hit is known.[67] Prefractionation (i.e., pladienolide B/nuclear extracts[106]) increases the concentration of the molecular target(s), and reduces the amount and the number of proteins in the lysate.

Washing cycles with increasing solubilization power are used to remove most aspecifically bound proteins before denaturation, if the FCG hit- target interactions are strong. Immobilized proteins are then eluted using denaturing solutions that disrupt the interactions between the supported FCG hit and the target(s),[107] or by concentrated solutions of free FCG hit (e.g., FK-506[108]). Selectively cleavable linkers with labile chemical groups (i.e., diazobenzenes-reducing conditions,[109] o-nitroarenes-photolabile linkers[110]) allow the release of FCG hit-specific targets without affecting the aspecific binding of other proteins to the matrix.

Target proteins are separated from aspecific protein binders in various ways. The biological extract can be passed through a supported FCG hit (step 1b, path A, Fig. 2.9) and through a structurally close, inactive analogue supported onto the same linker support (step 1a, e.g., aurilide/active and 6-epi-aurilide/inactive[111]). A comparison between eluates identifies missing FCG hit-specific targets from the inactive analogue elution (green target, path A, Fig. 2.9).

An excess of free FCG hit can be added to the cell lysate before its loading onto the supported FCG hit (step 6, path B, e.g., CB30865[112]). The soluble FCG hit binds to the target, and prevents its binding to the support. A comparison between eluates highlights the absence of FCG hit-specific targets from the soluble FCG hit-treated column (green target, path B).

The cell lysate can be submitted to serial affinity chromatography runs onto the supported FCG hit (step 7, path C, e.g., methotrexate[108]). A comparison between absorption-elution cycles should discriminate between target-hit interactions (lower amount of green target in second run eluate) and aspecific protein-hit interactions (same amounts in both eluates, path C, Fig. 2.9). Different paths are a better fit to different projects. An inactive, *quasi*-identical

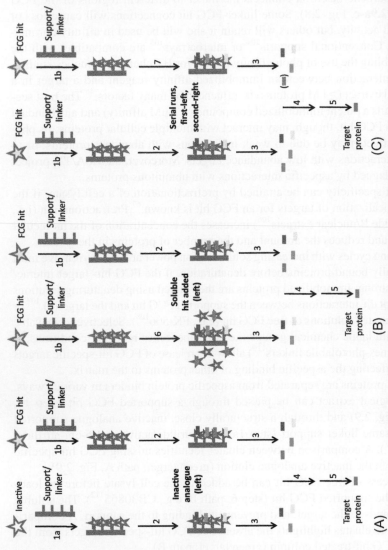

1a: immobilization of an inactive analogue, loading into an affinity column; 1b: as 1a, with FCG hit; 2: loading of a cell lysate, washings; 3: elution of bound protein targets; 4: SDS-PAGE separation of protein targets; 5: MS structure deconvolution; 6: as 2, adding soluble FCG hit; 7: loading of cell lysates from first run.

FIG. 2.9 Specific vs. aspecific interactions of immobilized affinity reagent-protein: methods for their discrimination.

analogue may not be available, or could have different physicochemical properties, inducing diverse interactions with a cell lysate. A poorly soluble FCG hit may prevent its preincubation with the lysate, or its use as a competitive eluant.

Fast structure determination for the eluted proteins, even in complex mixtures, is obtained with chromatographic (protein separation[113]) and MS methods (MW determination[114]). Specific FCG hit-target interactions can be also determined through metabolic or chemical labeling of putative protein targets.[41]

Metabolic labeling is achieved through *stable isotope labeling by aminoacids in cell culture* (*SILAC*,[115] Fig. 2.10).

SILAC entails cell growth in a medium containing either standard Arg and Lys, or ^{13}C- and ^{15}N-containing Arg and Lys (step 1a-light medium and step 1b-heavy medium, respectively, Fig. 2.10). After several cell replication cycles cells are lysed (steps 2a and 2b). Heavy and light lysates are loaded onto the supported FCG hit (step 3b) and onto a supported inactive probe (step 3a), respectively. After washing, eluting and pooling the eluates (step 4), the pooled proteins are submitted to LC-MS-MS analysis (step 5). Each bound protein is detected as a heavy-light couple of peaks. Proteins, which bind the supports through aspecific interactions, show equivalent MS peaks (dark/*heavy* and lighter/*light* lines, Fig. 2.10). Target peaks are present only from heavy isotope-supplemented extracts, and as single *blue/heavy* peaks in MS (Fig. 2.10). SILAC was applied to the identification/validation of targets for a variety of SMs.[116]

Alternatively, chemical labeling can be achieved on lysates through *isotope-coded affinity tag* (*ICAT*,[117] Fig. 2.11).

Cell lysates are loaded onto an immobilized FCG hit (step 1a, Fig. 2.11) or onto a supported, inactive analogue (step 1b). After washing and elution (step 2), the eluates are treated with a heavy/X=D and a light/X=H biotinylated affinity

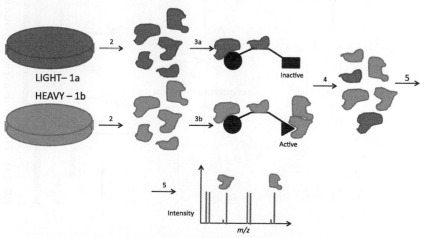

1a: growth, Arg-Lys medium; 1b: growth, ^{13}C/^{15}N Arg-Lys medium; 2: cell lysis; 3a: loading onto supported inactive probe; 3b: loading onto supported FCG hit, washings; 4: eluting and pooling bound proteins; 5: LC-MS-MS-based structure deconvolution.

FIG. 2.10 Metabolic labeling of target proteins in cells: SILAC.

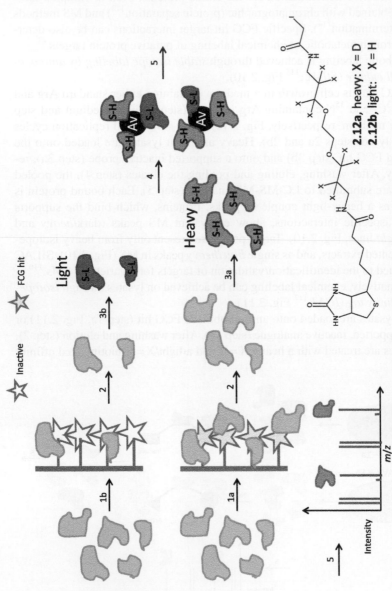

1a: immobilization, supported FCG hit; 1b: immobilization, supported inactive compound; 2: washing, elution; 3a: incubation with heavy 2.12a; 3b: incubation with light 2.12b; 4: immobilization, avidin beads; 5: LC-MS-MS-based structure deconvolution.

FIG. 2.11 Chemical labeling in cell lysates: ICAT.

label (respectively step 3a/**2.10a** and step 3b/**2.10b**). The iodoacetate in **2.10a,b** ensures labeling of each Cys-containing protein (estimated at ≈96% of the proteome) in both eluates. After pooling and tryptic digestion, the eluates are treated with avidin beads (step 4). The biotin group in protein-bound **2.10a,b** ensures their interaction with supported avidin. Avidin beads are then eluted and submitted to LC–MS–MS-based structure determination (step 5). Each protein is detected as a heavy (MW + 8) – light (MW) couple of peaks. Proteins, which bind the support through aspecific interactions, show equivalent MS peaks (dark/ *heavy* and lighter/*light* lines, Fig. 2.11), while specific target peaks are mostly present in heavy isotope/MW+8-supplemented extracts (Fig. 2.11). A more efficient ICAT version entails ^{13}C/heavy- and ^{12}C/light-containing affinity labels.[118]

Isobaric relative and absolute tag for quantification (iTRAQ[119]) tags proteins in a Cys-independent manner, and (albeit being operationally more complex) covers the entire proteome.

Soluble *matrix-free affinity reagents*[101,120] contain a tag for hit- target detection. A fluorophore tag can be used (e.g., cyclosporine A and FK-506[121]) to label an FCG hit-interacting target, although a noncovalent interaction often does not survive the procedures for target purification and isolation. A biotin tag,[122] after incubation of the labeled FCG hit with a biological sample, strongly interacts with avidin beads and purifies the targets. *Trifunctional probes* (e.g., duocarmicin A[123]) bear a fluorophore (detection tag) and biotin (enrichment tag) to identify (fluorophore) and enrich (biotin) molecular targets.

A bulky tag may alter the properties of an active compound, either by reducing its activity, or by modifying its physicochemical properties. *Bio-orthogonal probes*[124,125] are based on biocompatible reactions that take place in living cells without affecting them. A small reactive group is grafted onto the FCG hit prior to its incubation with a cell line, then a cell-permeable bulky tag containing a complementary reactive group is added to the same cells. A bio-orthogonal reaction forms a tag-FCG hit-target complex that is detected after cell lysis by chemical proteomics, or in living cells by imaging.[124] Live cell imaging shows the cellular localization of bio-orthogonal probes, and enables extract prefractionation to simplify the procedures for target isolation and deconvolution. It allows cellular visualization of protein targets with low stability that may not stand cell lysis.

Bio-orthogonal reactions include Cu(I)-catalyzed Huysgen cycloaddition/ click chemistry[126] of azides with alkynes (bioactive alkyne compound **2.11**; rhodamine B-based azide tag **2.12**; and bio-orthogonal probe **2.13**,[127] Fig. 2.12); Staudinger ligation[128] between methylester-containing triphenylphosphines and azides (bioactive azide compound **2.14**; FLAG peptide epitope DYKDDDDK-based triphenylphosphine tag **2.15**; and bio-orthogonal probe **2.16**[129]); and tetrazine ligation/inverse electron demand Diels Alder (IEDDA)[130] between tetrazines and *trans*-cyclooctene, or other strained alkenes (bioactive *trans*-cyclooctene compound **2.17**; rhodamine B-based tetrazine tag **2.18**; and bio-orthogonal probe **2.19**,[131] Fig. 2.12).

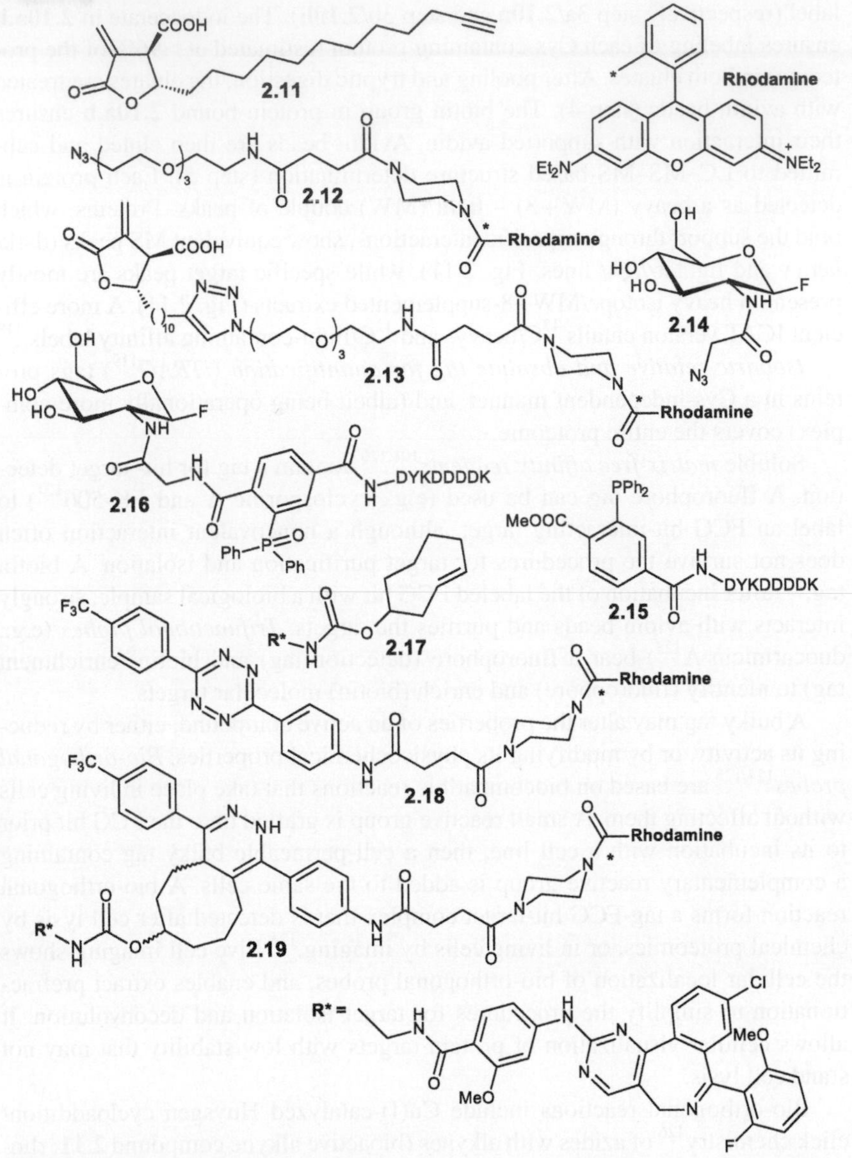

FIG. 2.12 Bio-orthogonal reactions and probes: click chemistry, Staudinger ligation, and tetrazine ligation/IEDDA.

Probes **2.13** (fatty acid synthase/FASN and carnitine O-palmitoyltransferase-1A/CPT-1A[127]), **2.16** (exoglycosidase specificity[129]) and **2.19** (nuclear distribution C/NUDC and other secondary targets[131]) enabled the identification of putative targets for various therapeutic indications.

Weak and reversible probe/target interactions may be sensitive to the processing of biological samples. A chemical probe covalently bound to its target (*activity-based protein profiling (ABPP)*)[132,133] answers this issue. A covalent bond connects an FCG hit-derived probe to its target through a reactive group on the probe that reacts with an aminoacidic residue in the target binding site. FDA-approved ibrutinib (**2.20**, Fig. 2.13) covalently binds to the Cys481

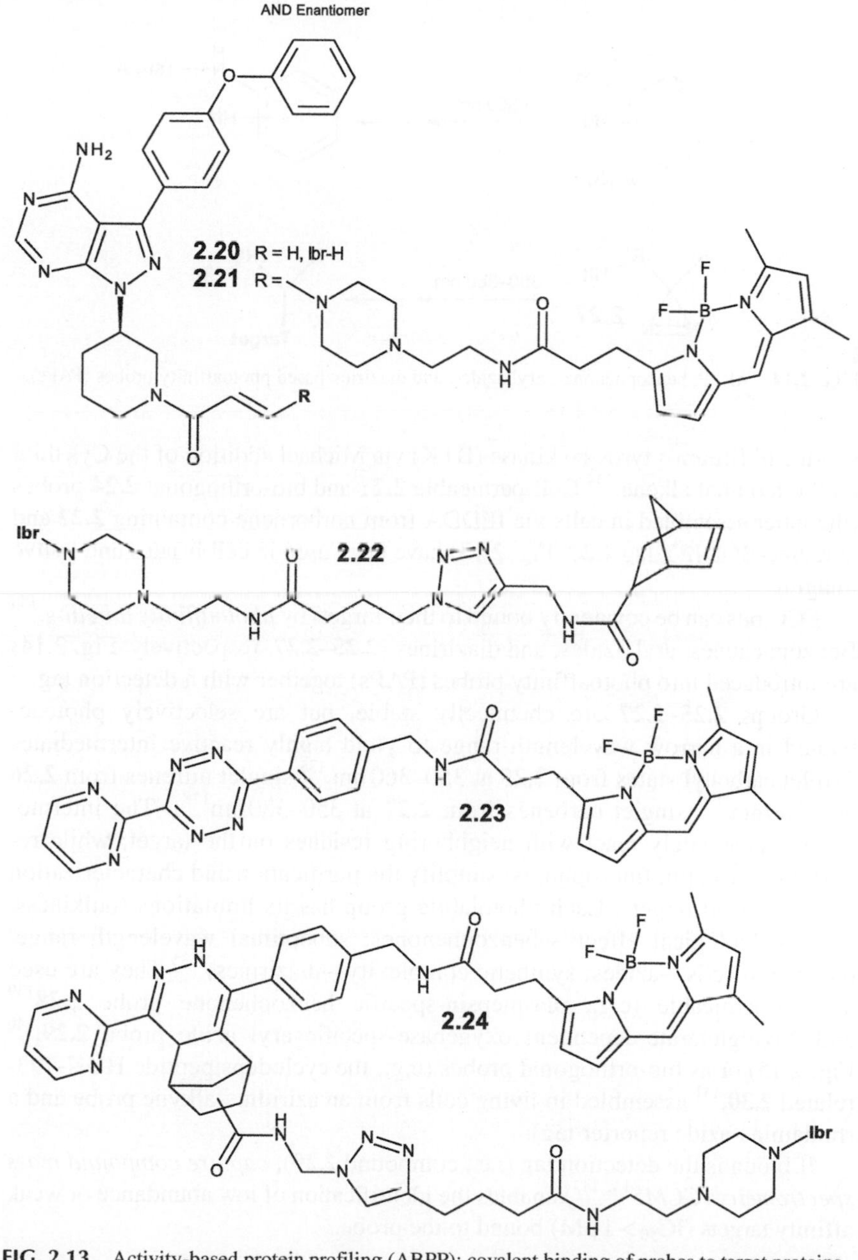

FIG. 2.13   Activity-based protein profiling (ABPP): covalent binding of probes to target proteins.

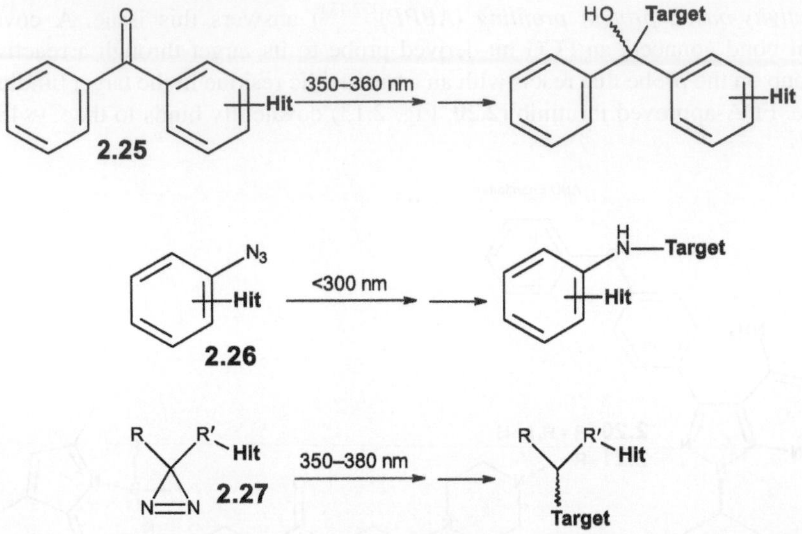

FIG. 2.14 ABPP: benzophenone-, aryl azide-, and diazirine-based photoaffinity probes (PAPs).

residue of Bruton's tyrosine kinase (BTK) via Michael addition of the Cys thiol on the terminal alkene.[134] Cell-permeable **2.21** and bio-orthogonal **2.24** probes (the latter assembled in cells via IEDDA from norbornene-containing **2.22** and tetrazine-BODIPY tag **2.23**, Fig. 2.13) have been used in cell lysates and in live imaging.[134]

FCG hits can be covalently bound to their targets by *photoaffinity labeling*.[135] Benzophenones, aryl azides, and diazirines (**2.25–2.27**, respectively, Fig. 2.14) are introduced into photoaffinity probes (PAPs) together with a detection tag.

Groups **2.25–2.27** are chemically stable, but are selectively photoactivated in a narrow wavelength range to yield highly reactive intermediates (triplet carbonyl states from **2.25** at 350–360 nm,[136] singlet nitrenes from **2.26** at <300 nm,[137] singlet carbenes from **2.27** at 350–380 nm[138]). The intermediates immediately react with neighboring residues on the target, while reporter tags (biotin, fluorophores) simplify the purification and characterization of PAP-bound targets. Each photolabile group has its limitations (bulkiness/altering biological effects—benzophenones; suboptimal wavelength range/damage to cells—azides; synthetic complexity—diazirines).[135] They are used as cell-permeable (e.g., plasmepsin-specific benzophenone probe **2.28**[139] and 2-oxoglutarate-dependent oxygenase-specific aryl azide probe **2.29**,[140] Fig. 2.15) or as bio-orthogonal probes (e.g., the cyclodepsipeptide HUN-793-related **2.30**,[141] assembled in living cells from an aziridine-alkyne probe and a rhodamine-azide reporter tag).

If biotin is the detection tag (i.e., compound **2.29**), *capture compound mass spectrometry (CCMS*[142,143]) enables the identification of low abundance or weak affinity targets (IC$_{50}$ > 1 μM) bound to the probe.

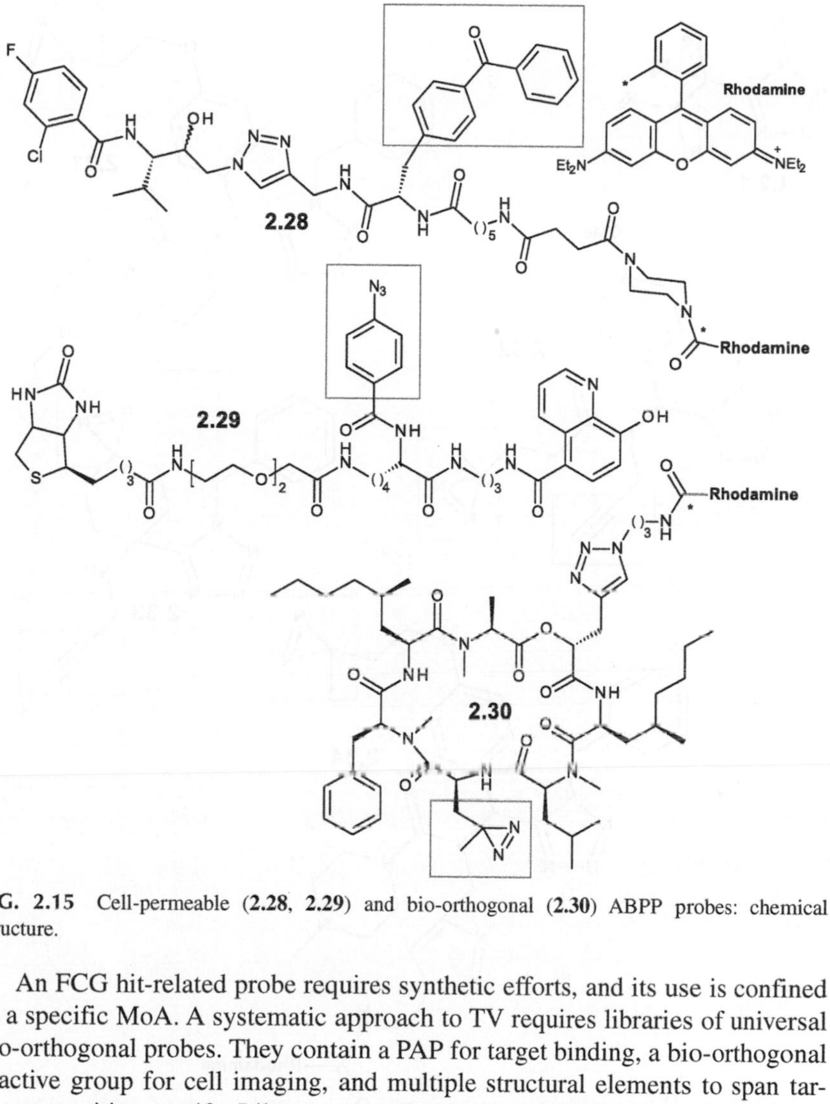

FIG. 2.15 Cell-permeable (**2.28**, **2.29**) and bio-orthogonal (**2.30**) ABPP probes: chemical structure.

An FCG hit-related probe requires synthetic efforts, and its use is confined to a specific MoA. A systematic approach to TV requires libraries of universal bio-orthogonal probes. They contain a PAP for target binding, a bio-orthogonal reactive group for cell imaging, and multiple structural elements to span target recognition motifs. Library screening on biological extracts should identify multiple probe-protein interactions; could determine the druggability of putative target classes; and should support the acquisition of SARs for such targets. The 60-membered tetrazole library **L2.1**[144] (Fig. 2.16) was prepared by Ugi azide multicomponent condensation between isocyanides, azides, and amines. Each library individual (e.g., **2.31–2.33**) carried a diazirine PAP and a bio-orthogonal triple bond, and could react with a rhodamine-azide tag (**2.34** from **2.31**, Fig. 2.16) to yield a bio-orthogonal probe.

Putative targets were identified after **L2.1** testing in PC-3 prostate cancer cells, using SILAC labeling and enrichment. Probe specificity was observed for

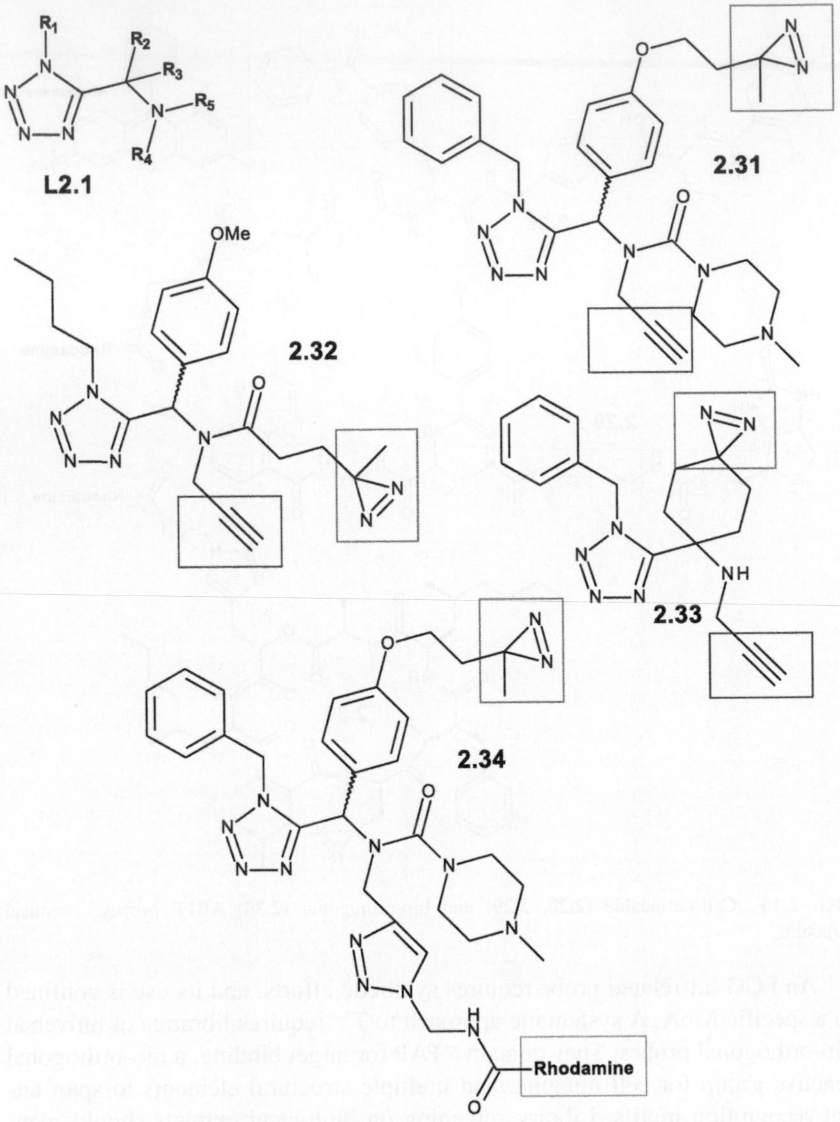

FIG. 2.16 Universal bio-orthogonal probes: structure of a library (**L2.1**) and of some target-specific hits/probes (**2.31–2.34**).

druggable targets (**2.31**- nucleoside diphosphate linked moiety X-type 1 motif/NUDT1, a phosphatase; **2.32**-phosphatidyl ethanolamine binding protein 1/ PEBP1, a regulator of Raf kinase), and for unprecedented targets (**2.33**-breast cancer antiestrogen resistance 3/BCAR3, an adaptor protein).[144]

Expression-cloning techniques are useful with low-abundance target protein(s).[67,120] *Phage display* (Fig. 2.17, top)[145] requires the insertion of a

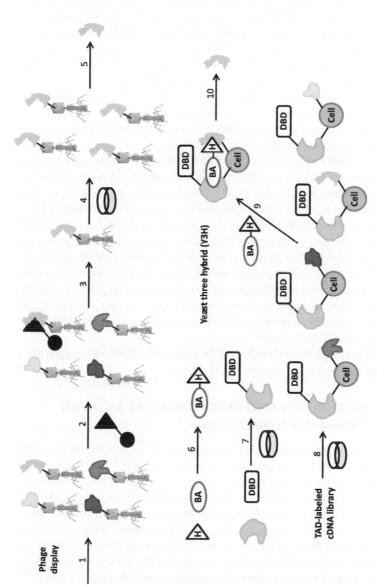

1: cDNA library transfection; **2**: incubation with immobilized FCG hit; **3**: washings and elution of bound phages; **4**: phage amplification / biopanning in *E. coli*; **5**: phage DNA sequencing; **6**: hit-bait fusion; **7**: EA receptor-DBD fusion; **8**: TAD-labeled cDNA library transfection; **9**: BA-H screening, reporter gene expression, selection of target-bound cells; **10**: cell lysis, target deconvolution.

FIG. 2.17   Expression-cloning techniques in TV: phage display, yeast three hybrids (Y3H).

cDNA gene library into a population of phages. The resulting phage library displays proteins fused onto phage coat proteins (step 1). Screening of the phage library with an immobilized FCG hit selects phages displaying putative targets (step 2). After washing and elution of target-bound phages (step 3), selected phages are transfected and amplified in host cells (e.g., *E.coli*, biopanning, step 4). Their DNA is sequenced, and the genes encoding for FCG hit-targets are deconvoluted (step 5, Fig. 2.17, top). Phage display does not ensure proper folding or PTMs for displayed proteins.[67] It was used to elucidate the MoA of several SMs (e.g., kahalalide F—human ribosomal protein S25[146]).

Yeast three-hybrid systems (Y3H, Fig. 2.17, bottom)[147] require the fusion of an FCG hit (H) with an active known molecule (bait, BA) in a hybrid (step 6). The hybrid retains the affinity of both molecules for their targets. The BA receptor (e.g., methotrexate-dehydrofolate reductase[148]) is fused to a DNA-binding domain/DBD in a suitable yeast strain (step 7). Yeast cells are then transfected with a cDNA library that encodes proteins fused with a transcriptional activation domain/TAD (step 8). Then, screening for putative targets with the DBD-BA receptor-BA-H hybrid looks for the interaction of H with a TAD-fused protein (step 9). The H-bound TAD domain close to the BA-bound DBD promotes the expression of a reporter gene and the selection of the yeast cells expressing the reporter (step 9). Cell lysis and TAD-labeled target deconvolution (step 10, Fig. 2.17, *bottom*) identifies the target. A good expression of the reporter gene ensures the identification of low-abundance targets. Conversely, the hybrid molecule must be cell permeable and stable, and proper folding and PTMs are valid concerns. Examples of Y3H-assisted TI using SMs are reported (e.g., antituberculosis drug targets[149]).

The next Paragraph describes the TV of a class of "difficult" targets using FCG hits, and structurally related chemical probes.

2.3. BROMODOMAIN AND EXTRATERMINAL DOMAIN (BET) READERS—I-BET (GSK525762A)

The cell nucleus of an eukaryotic cell has a ≈ 0.01-mm diameter, and a ≈ 1-pL volume. The stretched human genome, composed of 6×10^9 base pairs, would be ≈ 2-m long.[150] DNA packing into *chromatin*[151] takes place through five sub-families of *histone proteins* (H1, H2A, H2B, H3, and H4,[152] top, Fig. 2.18). Two copies of H2A, H2B, H3, and H4 assemble into a symmetrical octamer spool (*light* cylinder). ≈ 147 DNA base pairs (bps, thick black lines) wrap ≈ 1.65 times around each octamer forming a *nucleosome*.[153] Wrapping on the H2A-H4 octamer takes place with any DNA, and is promoted by the attraction between basic/positively charged histone proteins and acidic/negatively charged DNA.[154]

A linker DNA segment (20 to 80 bps) connects nucleosomes, an H1A copy locks the incoming and outcoming DNA fibers on each nucleosome, and a "beads on a string" chromatin structure is formed (≈ 11-nm diameter).[154] Up to 30% of H2A-H4 proteins is made by N-terminal disordered regions, named

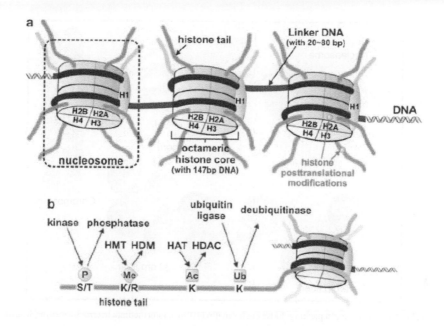

Figure 1, pag. 1239

Cell Death and Differentiation (2010) **17**, 1238–1243; doi:10.1038/cdd.2010.58;

FIG. 2.18   Chromatin structure: histone proteins, DNA, nucleosomes, histone tails, PTMs. *From Fig. 1, page 1239. Cell Death and Differentiation (2010) 17, 1238-1243. doi: 10.1038/cdd.2010.58.*

histone tails (thick protruding *lines*, Fig. 2.18A).[155] Histone tails are Lys- and Arg-rich, positively charged sequences that fluctuate away from the nucleosome core and experience dynamic PTM patterns (*bottom*, Fig. 2.18B).

The loose arrangement of "beads on a string" chromatin (bottom left, Fig. 2.19) enables DNA transcription.[156] Intermediate packing arrangements (e.g., a≈30-nm diameter, 180-mm-length structure,[157] bottom right, Fig. 2.19) give limited chromatin access to the DNA transcription machinery.[157] The equilibrium between loose (euchromatin) and intermediate packing (heterochromatin) of varying lengths regulates gene expression in nonreplicating cells.[158]

Further condensation of heterochromatin during mitosis leads to densely packed chromosomal DNA (≈700-nm diameter, ≈120-mm length, *top right*, Fig. 2.19). Chromosomes prevent DNA transcription and protect DNA integrity during cell replication.[158]

The dynamic nature of chromatin packing and gene expression depends on PTM patterns on histone tails[159] and on the lateral nucleosome surface.[160] *Methylation* (mono-, bis-, and tri-, Lys and Arg), *acetylation* (Lys, Arg, Ser, Thr, and Tyr),[161] and *phosphorylation* (Ser, Thr, and Tyr)[162] of N-terminal H2A, H2B, H3, and H4 proteins were the first observed PTMs. Modern technologies

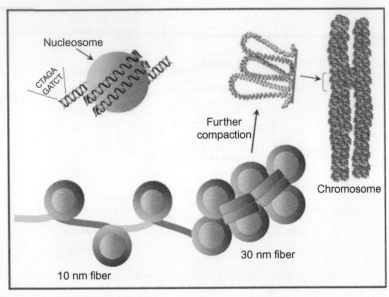

FIG. 2.19 Chromatin packing: loose euchromatin (10 nm), intermediate heterochromatin (30 nm), dense chromosomal DNA.

detected low abundance PTMs (e.g., ubiquitination, SUMOylation, acylation, hydroxylation, and glycosylation[159]) that also influence chromatin packing and gene expression.

Several hundreds of histone PTMs on ≈100 histone residues were observed in vitro, in human or rodent samples.[163] Some reduce the basic/positively charged nature of histones (acylation of Lys and Arg residues/loss of positive charge; phosphorylation of Ser, Thr, and Tyr residues/introduction of a negative charge). Then, histone octamers decrease their affinity for negatively charged DNA, and the resulting loose chromatin is more accessible to the transcription machinery.[164]

Charge-neutral Lys and Arg methylation is the most abundant and site-selective histone modification.[165] A few histone PTMs are solidly connected with the activation of gene expression (e.g., trimethylated Lys4 on H3/H3K4me3,[166] monomethylated Lys20 on H4/H4K20me,[167] and acetylated Lys9 on H3/H3K9ac[168]), with its repression (e.g., H3K27me3,[169] and H3K9me2[170]), or with mitotic chromosome condensation (e.g., phosphorylated Ser10 on H3/H3S10ph[171]). Multiple PTMs are observed on histones, including bivalent activator/H3K4me3-repressor/H3K27m3 marks[172] and large patterns, containing up to 17 colocalizing histone modifications.[173]

Histone PTMs are also named *histone marks*. Their presence, or absence, tags the histone proteins (mostly on N-terminus) and promotes, or prevents, the interaction with chromatin-remodeling enzymes. Chromatin-remodeling enzymes induce the removal or the addition of further histone marks.[174] The complexity of

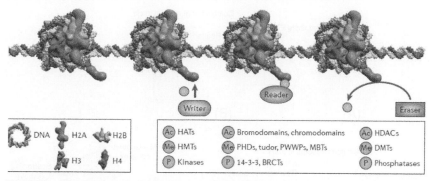

FIG. 2.20   Histone marks and chromatin-remodeling enzymes: writers, erasers, and readers.

histone PTMs and their recurrence led to the *histone code* hypothesis[175]—combinations of histone PTMs regulating the transcription of DNA. The code includes novel histone marks and PTM categories, interhistone marks patterns, histone variants, and so on. Fig. 2.20 shows histone marks and chromatin-remodeling enzymes.

Histone marks are introduced by *writer enzymes* (histone methyltransferases/HMT,[176] histone acetyltransferases/HAT,[177] and kinases[178]), and removed by *eraser enzymes* (histone demethylases/HDM,[179] histone deacetylases/HDAC,[180] and phosphatases[181]). Their presence usually promotes the docking of site- and PTM-specific *histone readers*[182] (Fig. 2.20). Among them, bromodomains[183] were discovered as Lys acetylation readers. Lys and Arg methylation readers include plant homeodomains (PHDs[184]), chromodomains (CHDs[185]), and tandem Tudor domains (TTDs[186]). The 14-3-3 protein[187] contains a Ser phosphorylation reader domain.

Histone readers may selectively bind to two adjacent histone marks through their recognition domain (path A, Fig. 2.21). Two (or more) marks and reader domains may be required to establish a combinatorial *cis*-recognition in the same histone tail (path B), a *trans*-recognition in two (or more) neighboring tails of the same nucleosome (path C-left), or of different nucleosomes (path C-right).[182]

Readers, writers, and erasers participate to complex recognition/histone modification/chromatin-remodeling events, where scaffolding proteins assemble multisubunit complexes with multiple activities impacting on DNA transcription. A graphical representation is depicted in path D, Fig. 2.21.[182]

DNA methylation,[188] nucleosomes bearing histone protein variants,[189] and nuclear compartmentalization[190] add complexity to the organization and functions of chromatin. The long half-life of methylated histones (≈ 1 day) indicates its relevance in the transmission of stabilized epigenetic traits to next generations of cells.[191] Short-lived acetylated and phosphorylated histones (≤ 30 min for ac, up to 2 h for ph) modulate chromatin remodeling in cells with limited (if any) epigenetic transmission.[191]

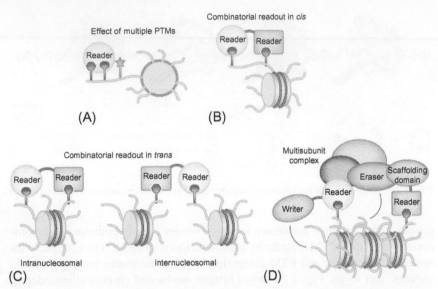

FIG. 2.21 Histone marks and chromatin-remodeling enzymes: interaction modes, protein partners.

Further investigations are needed to translate our knowledge into drugs, although HDACs (marketed drugs, strong clinical pipeline[192]), bromodomains (phase II clinical trials[193]), HDMs (phase I clinical trials[194]), HMTs[195] and HATs[196] (preclinical studies) are validated writer, eraser, or reader targets. FCG studies and chemistry-driven TV were instrumental in validating histone PTM-related targets.

The interaction of readers and their chromatin partners through a histone mark pattern was considered relevant, but hardly druggable.[197] The *bromodomain and extraterminal domain (BET[193])* subfamily of bromodomain readers was known to regulate proinflammatory gene expression,[198,199] to repress Tat-mediated transactivation of the HIV promoter,[200] and to deregulate proteins in cancer.[201] Prior to 2010, BET modulators (either SMs, or biologicals) were unknown.

Researchers from GSK and the Rockefeller University described a phenotypic assay,[202,203] based on a human HepG2 hepatocyte cell line containing an apolipoprotein A-1 (ApoA1) luciferase reporter gene. Activators of ApoA1 were targeted for their anti-inflammatory activity that may protect from atherosclerosis.[204] An FCG screen, run on a chemical library (unknown size), identified SM upregulators of reporter gene activity[202] (step 1, Fig. 2.22).

Benzodiazepine **2.35** (Fig. 2.23) was identified as an ApoA1 inducer with submicromolar activity ($EC_{50} = 440$ nM, maximum induction ≈ 4.5-fold vs. control levels).[203] Its induction was ApoA1-specific, as it was inactive on HepG2 cells containing an LDL-R promoter reporter (step 2, Fig. 2.22). The synthesis of a few analogues (step 3, Fig. 2.22) established a preliminary SAR.

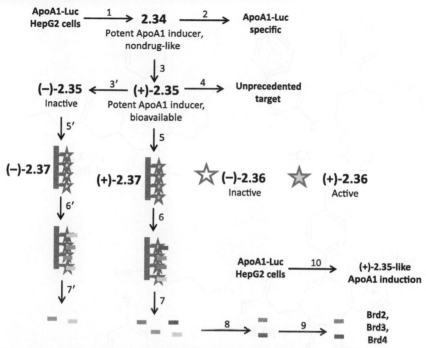

FIG. 2.22   Identification of the selective BET reader inhibitor I-BET: screening cascade, TI and TV studies.

Enantiomerically pure (+)-**2.36** (I-BET, Fig. 2.23) coupled the induction of ApoA1 with good bioavailability and metabolic stability, and was selected for MoA studies. Its enantiomer (−)-**2.36** was completely inactive in the FCG assay (step 3′, Fig. 2.22).[203]

I-BET was inactive in vitro on kinases, cytochrome isoforms, ion channels, GPCRs, and transporter proteins (step 4, Fig. 2.22).[202] An aminopropane linker was added onto **2.36** enantiomers (only (+)-**2.37** shown in Fig. 2.23). Their free amino group was anchored onto agarose gel (steps 5 and 5′), and the resulting immobilized (+)-**2.37** and (−)-**2.37** were treated with extracts from HepG2 cells (steps 6 and 6′). After washing of unbound proteins and putative target elution (steps 7 and 7′), the comparison between eluates identified several (+)-**2.37**—specific target sequences (step 8). The sequences were deconvoluted by LC/MS/MS (step 9), corresponding to full-length (FL) Brd2, Brd3, and Brd4 proteins (three BET family members).[203] Brd4-specific siRNAs recapitulated the ApoA1 induction phenotype caused by (+)-**2.36** (step 10, Fig. 2.22).

The identification of (+)-**2.36** (I-BET), and of a similar, BET-specific probe ((+)-**2.38**, JQ-1, Fig. 2.23)[205] validated BET readers as druggable targets for therapeutic applications.

Affinity chromatography using HepG2 cells transfected with Brd constructs identified their N-terminal region (containing two bromodomains) as

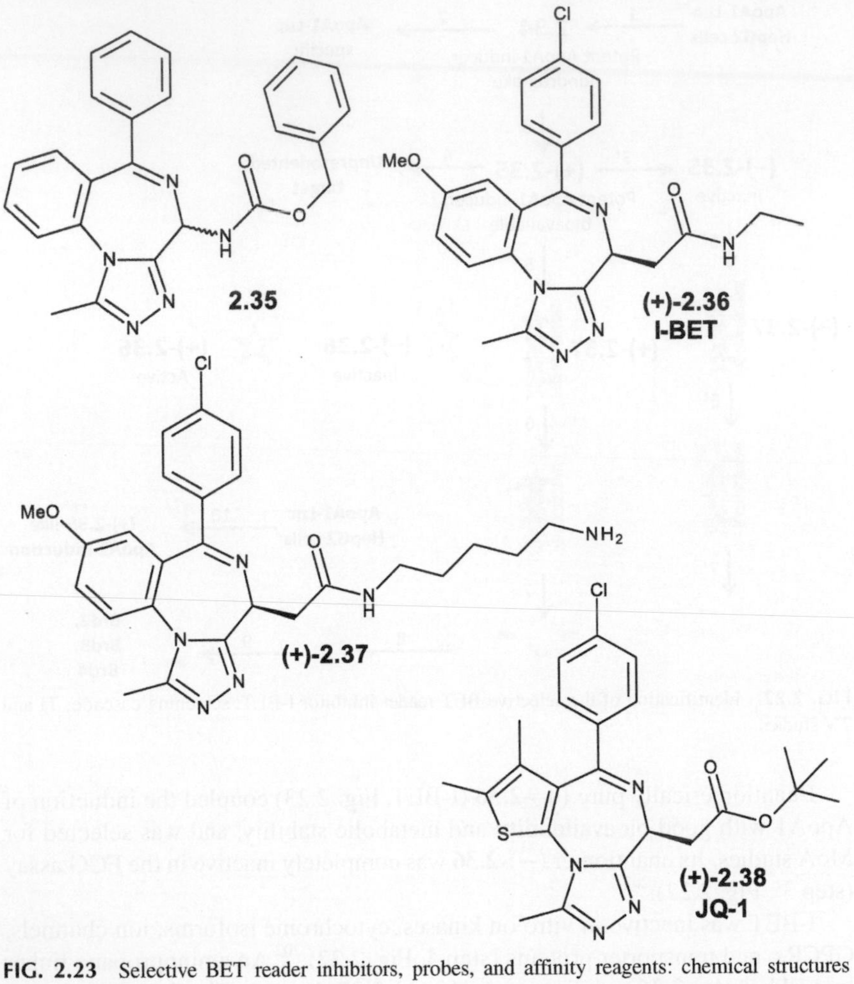

FIG. 2.23 Selective BET reader inhibitors, probes, and affinity reagents: chemical structures (2.35-(+)-2.38).

the binding site for (+)-2.36. Isothermal titration calorimetry (ITC, (+)-2.36 and (+)-2.38), differential scanning calorimetry (DSC, (+)-2.38), and surface plasmon resonance (SPR, (+)-2.36) determined binding affinities for both bromodomains in the 10–200 nM range.[203,205] A 1:2 Brd:probe stoichiometry was observed with constructs containing two bromodomains. Faster binding kinetics were observed for the first bromodomain, suggesting multistep/cooperative binding events of each domain with its acetylated histone marks.[203]

The binding mode was determined from the X-ray complexes between inhibitors and bromodomains of Brd2 and Brd4. Similar binding modes for (+)-2.36, (+)-2.38 and ischemin[206] (bottom (C), top right (B), and top left (A), respectively) with the bromodomains of Brd4 are shown in Fig. 2.24.[207]

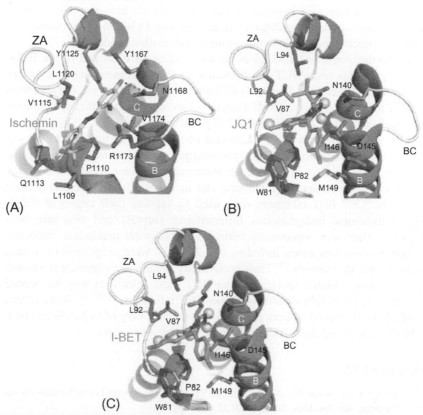

FIG. 2.24 X-ray complexes of Brd proteins with BET reader inhibitors: ischemin (top *left*), Brd4-(+)-**2.38** (top right), Brd4-(+)-**2.36** (bottom).

The compounds establish key interactions with a conserved Asn residue, and fill the Kac-binding site preventing binding of acetylated histone residues.[207] (+)-**2.36** and (+)-**2.38** are inactive against bromodomains from non-BET reader proteins, due to structural differences. They can be classified as acetyllysine mimetic, BET-specific inhibitors.[207]

Biological profiling in inflammation ((+)-**2.36**[202]) and cancer models ((+)-**2.38**[205]) validated BET family members in these therapeutic areas. Nonstimulated bone-marrow-derived macrophages (BMDMs) treated with (+)-**2.36** showed marginal effects on gene expression (lack of aspecific/toxic effects, including housekeeping genes). Lipopolysaccharide (LPS)-stimulated and tumor necrosis factor (TNF)-stimulated BMDMs treated with (+)-**2.36** showed suppression of a panel of inflammation-induced genes.[202] Most unaffected genes were early response/primary stimulated genes, with a high acetylation/methylation content on histones (e.g., histone mark-driven stimulation preceding treatment with (+)-**2.36**). Most suppressed genes were late response/secondary stimulated genes, with a low acetylation/methylation content on histones (e.g., histone mark-driven stimulation inhibited by treatment with (+)-**2.36**). Compound (+)-**2.36** was effective in vivo in three murine inflammation models.[202]

As to oncology, (+)-**2.38** inhibited the binding of Brd4 and of the Brd4-nuclear protein in testis/NUT oncoprotein to nuclear chromatin in human osteosarcoma (U2OS) cells.[202] It induced differentiation, growth arrest, and apoptosis in NUT midline carcinoma (NMC) cells, and showed in vivo efficacy in murine xenografts of patient-derived Brd4-NUT-positive NMCs.[202]

The characterization of (+)-**2.36** and (+)-**2.38** paved the way to the exploitation of BET inhibitors. Other chemotypes were identified as acetyllysine mimetic, BET-specific bromodomain inhibitors; as nonacetyllysine mimetic, BET-specific bromodomain inhibitors; and as nonBET-specific bromodomain inhibitors.[207–209] BET inhibitors were used to validate their targets in oncology (hematologic malignancies), inflammation (sepsis), and viral infections (HIV).[210] They were structurally optimized, underwent preclinical characterization and—in four cases, including (+)-**2.36**[211]—were progressed to clinical trials, mostly in oncology.[193] Their preliminary clinical development is encouraging, although higher specificity/inter-BET class selectivity may be needed to overcome potential side effects of pan-BET inhibitors.[193,212,213] None of this would have happened without the FCG-driven discovery of (+)-**2.36** and (+)-**2.38** that elucidated the role of BET readers on histone marks.

REFERENCES

1. Plenge, R. M.; Scolnick, E. M.; Altshuler, D. Validating Therapeutic Targets Through Human Genetics. *Nat. Rev. Drug Discov.* **2013**, *12*, 581–594.

2. DiMasi, J. A.; Faden, L. B. Competitiveness in Follow-on Drug R&D: A Race or Imitation? *Nat. Rev. Drug Discov.* **2011**, *10*, 23–27.

3. http://www.fda.gov/downloads/drugs/guidancecomplianceregulatoryinformation/guidances/ucm358301.pdf.

4. http://csdd.tufts.edu/files/uploads/PubPrivPaper2015.pdf.

5. Paul, S. M.; Mytelka, D. S.; Dunwiddie, C. T.; Persinger, C. C.; Munos, B. H.; Lindborg, S. R.; Schacht, A. L. How to Improve R&D Productivity: The Pharmaceutical industry's Grand Challenge. *Nat. Rev. Drug Discov.* **2010**, *9*, 203–214.

6. Calcoen, D.; Elias, L.; Yu, X. What Does it Take to Produce a Breakthrough Drug? *Nat. Rev. Drug Discov.* **2015**, *14*, 161–162.

7. Hay, M.; Thomas, D. W.; Craighead, J. L.; Economides, C.; Rosenthal, J. Clinical Development Success Rates for Investigational Drugs. *Nat. Biotech.* **2014**, *32*, 40–51.

8. Arrowsmith, J.; Miller, P. Phase II and Phase III Attrition Rates 2011–2012. *Nat. Rev. Drug Discov.* **2013**, *12*, 569.

9. Prinz, F.; Schlange, T.; Asadullah, K. Believe it or Not: How Much Can We Rely on Published Data on Potential Drug Targets? *Nat. Rev. Drug Discov.* **2011**, *10*, 712.

10. Begley, C. G.; Ellis, L. M. Raise Standards for Preclinical Cancer Research. *Nature* **2012**, 531–533.

11. Haibe-Kains, B.; El-Hachem, N.; Birkbak, N. J.; Jin, A. C.; Beck, A. H.; Aerts, H.J.W.L.; Quackenbush, J. Inconsistency in Large Pharmacogenomics Studies. *Nature* **2013**, 389–393.

12. Hench, P. S.; Kendall, E. C.; Slocumb, C. H.; Polley, H. F. Effects of Cortisone Acetate and Pituitary ACTH on Rheumatoid Arthritis, Rheumatic Fever and Certain Other Conditions. *Arch. Intern. Med.* **1950**, *85*, 545–666.

13. Stone, E. An Account of the Success of the Bark of the Willow in the Cure of Agues. In a Letter to the Right Honourable George Earl of Macclesfield, President of R. S. From the Rev. Mr. Edmund Stone, of Chipping-Norton in Oxfordshire. *Phil. Trans. Royal Soc. London* **1763**, *53*, 195–200.

14. Dawber, T. R.; Meadors, G. F.; Moore, F. E., Jr. Epidemiological Approaches to Heart Disease: The Framingham Study. *Am. J. Public Health* **1951**, *41*, 279–286.

15. Rowley, J. D. Letter: A New Consistent Chromosomal Abnormality in Chronic Myelogenous Leukaemia Identified by Quinacrine Fluorescence and Giemsa Staining. *Nature* **1973**, *243*, 290–293.

16. Konopka, J.; Watanabe, S. B.; Witte, O. N. An Alteration of the Human c-abl Protein in K562 Leukemia Cells Unmasks Associated Tyrosine Kinase Activity. *Cell* **1984**, *37*, 1035–1042.

17. Daley, G.; Van Etten, R. A.; Baltimore, D. Induction of Chronic Myelogenous Leukemia in Mice by the P210bcr/abl Gene of the Philadelphia Chromosome. *Science* **1990**, *247*, 824–830.

18. Wang, D. G.; Fan, J. B.; Siao, C. J.; Berno, A.; Young, P.; Sapolsky, R.; Ghandour, G.; Perkins, N.; Winchester, E.; Spencer, J.; Kruglyak, L.; Stein, L.; Hsie, L.; Topaloglou, T.; Hubbell, E.; Robinson, E.; Mittmann, M.; Morris, M. S.; Shen, N.; Kilburn, D.; Rioux, J.; Nusbaum, C.; Rozen, S.; Hudson, T. J.; Lipshutz, R.; Chee, M.; Lander, E. S. Large-Scale Identification, Mapping, and Genotyping of Single-Nucleotide Polymorphisms in the Human Genome. *Science* **1998**, *280*, 1077–1082.

19. Cowper-Sal, R.; Cole, M. D.; Karagas, M. R.; Lupien, M.; Moore, J. H. Layers of Epistasis: Genome-Wide Regulatory Networks and Network Approaches to Genome Wide Association Studies. *Wiley Interdiscip. Rev. Syst. Biol. Med.* **2011**, *3*, 513–526.

20. Berg, E. L. Systems Biology In Drug Discovery and Development. *Drug Discov. Today* **2014**, *19*, 113–125.

21. Balasubramaniam, S.; Goldstein, J. L.; Brown, M. S. Regulation of Cholesterol Synthesis in Rat Adrenal Gland through Coordinate Control of 3-Hydroxy-3-Methylglutaryl Coenzyme a Synthase and Reductase Activities. *Proc. Natl. Acad. Sci.* **1977**, *74*, 1421–1425.

22. Brown, M. S.; Goldstein, J. L. Receptor-Mediated Control of Cholesterol Metabolism. *Science* **1976**, *191*, 150–154.

23. Bots, M. L.; Visseren, F. L.; Evans, G. W.; Riley, W. A.; Revkin, J. H.; Tegeler, C. H.; Shear, C. L.; Duggan, W. T.; Vicari, R. M.; Grobbee, D. E.; Kastelein, J. J.; RADIANCE 2 Investigators Torcetrapib and carotid intima-media thickness in mixed dyslipidaemia (RADIANCE 2 study): a randomised, double-blind trial. *Lancet* **2007**, *370*, 153–160.

24. Barter, P.; Rye, K. A. Cholesteryl Ester Transfer Protein Inhibition to Reduce Cardiovascular Risk: Where Are we Now? *Trends Pharmacol. Sci.* **2011**, *32*, 694–699.

25. Perrimon, N.; Friedman, A.; Mathey-Prevot, B.; Eggert, U. S. Drug-Target Identification in Drosophila Cells: Combining High-throughout RNAi and Small-Molecule Screens. *Drug Discov. Today* **2007**, *12*, 28–33.

26. Lawson, N. D.; Wolfe, S. A. Forward and Reverse Genetic Approaches for the Analysis of Vertebrate Development in the Zebrafish. *Dev. Cell.* **2011**, *21*, 48–64.

27. Denayer, T.; Stohr, T.; VanRoy, M. Animal Models in Translational Medicine: Validation and Prediction. *New Horiz. Translat. Med.* **2014**, *2*, 5–11.

28. McGonigle, P.; Ruggeri, B. Animal Models of Human Disease: Challenges in Enabling Translation. *Biochem. Pharmacol.* **2014**, *87*, 162–171.

29. Vidalin, O.; Muslmani, M.; Estienne, C.; Echchakir, H.; Abina, A. M. In Vivo Target Validation Using Gene Invalidation, RNA Interference and Protein Functional Knockout Models: It Is the Time to Combine. *Curr. Opin. Pharmacol.* **2009**, *6*, 669–676.

30. Doyle, A.; MP, M. G.; Lee, N. A.; Lee, J. J. The Construction of Transgenic and Gene Knockout/Knockin Mouse Models of Human Disease. *Transgenic Res.* **2012**, *21*, 327–349.

31. Moore, J. D. The Impact of CRISPR–Cas9 on Target Identification and Validation. *Drug Discov. Today* **2015**, *20*, 450–457.

32. Naftzger, C.; Takechi, Y.; Kohda, H.; Hara, I.; Vijayasaradhi, S.; Houghton, A. N. Immune Response to a Differentiation Antigen Induced by Altered Antigen: A Study of Tumor Rejection and Autoimmunity. *Proc. Natl. Acad. Sci. USA* **1996**, *93*, 14809–14814.

33. Silva, J.; Chang, K.; Hannon, G. J.; Rivas, F. V. RNA-Interference-Based Functional Genomics in Mammalian Cells: Reverse Genetics Coming of Age. *Oncogene* **2004**, *23*, 8401–8409.

34. Poste, G. Bring on the Biomarkers. *Nature* **2011**, *469*, 156–157.

35. Bunnage, M. E.; Piatnitski Chekler, E. L.; Jones, L. H. Target Validation Using Chemical Probes. *Nat. Chem. Biol.* **2013**, *9*, 195–199.

36. http://www.nobelprize.org/nobel_prizes/medicine/laureates/1950/.

37. Newman, D. J.; Cragg, G. M. Natural Products as Drugs and Leads to Drugs: The Historical Perspective. In *RSC Biomolecular Sciences No. 18; Natural Product Chemistry for Drug Discovery*; Buss, A. D., Butler, M. S., Eds.; RSC: Cambridge, UK, 2010; pp 3–27.

38. Munos, B. A Forensic Analysis of Drug Targets from 2000 Through 2012. *Clin. Pharmacol. Ther.* **2013**, *94*, 407–411.

39. Gregori-Puigjané, E.; Setola, V.; Hert, J.; Crews, B. A.; Irwin, J. J.; Lounkine, E.; Marnett, L.; Roth, B. L.; Shoichet, B. K. Identifying Mechanism-of-Action Targets for Drugs and Probes. *Proc. Natl. Acad. Sci. USA* **2012**, *109*, 11178–11183.

40. Strittmatter, S. M. Overcoming Drug Development Bottlenecks With Repurposing: Old Drugs Learn New Tricks. *Nat. Med.* **2014**, *20*, 590–591.

41. Schenone, M.; Dančík, V.; Wagner, B. K.; Clemons, P. A. Target Identification and Mechanism of Action in Chemical Biology and Drug Discovery. *Nat. Chem. Biol.* **2013**, *9*, 232–240.

42. Mayer, T. U.; Kapoor, T. M.; Haggarty, S. J.; King, R. W.; Schreiber, S. L.; Mitchison, T. J. Small Molecule Inhibitor of Mitotic Spindle Bipolarity Identified in a Phenotype-Based Screen. *Science* **1999**, *286*, 971–974.

43. Cereto-Massagué, A.; José Ojeda, M.; Valls, C.; Mulero, M.; Pujadas, G.; Garcia-Vallve, S. Tools for In Silico Target Fishing. *Methods* **2015**, *71*, 98–103.

44. Jacoby, E. Computational Chemogenomics. *WIREs Comput. Mol. Sci.* **2011**, *1*, 57–67.

45. Futamura, Y.; Muroi, M.; Osada, H. Target Identification of Small Molecules Based on Chemical Biology Approaches. *Mol. BioSyst.* **2013**, *9*, 897–914.

46. Cong, F.; Cheung, A. K.; Huang, S.-M.A. Chemical Genetics-Based Target Identification in Drug Discovery. *Annu. Rev. Pharmacol. Toxicol.* **2012**, *52*, 57–78.

47. Nikolova, N.; Jaworska, J. Approaches to Measure Chemical Similarity – A Review. *QSAR Combin. Sci.* **2003**, *22*, 1006–1026.

48. Willett, P. Similarity Searching Using 2D Structural Fingerprints. *Methods Mol. Biol.* **2011**, *672*, 133–158.

49. Nettles, J. H.; Jenkins, J. L.; Bender, A.; Deng, Z.; Davies, J. W.; Glick, M. Bridging Chemical and Biological Space: "Target Fishing" Using 2D and 3D Molecular Descriptors. *J. Med. Chem.* **2006**, *49*, 6802–6810.

50. Bajusz, D.; Racz, A.; Heberger, K. J. Why Is Tanimoto Index an Appropriate Choice for Fingerprint-Based Similarity Calculations? *J. Cheminform.* **2015**, *7*, 1–26.

51. Johnson, M. A., Maggiora, G. M., Eds. *Concepts and Applications of Molecular Similarity*; John Wiley & Sons: New York, 1990. 393 p.

52. Sidduri, A.; Tilley, J. W.; Lou, J. P.; Chen, L.; Kaplan, G.; Mennona, F.; Campbell, R.; Guthrie, R.; Huang, T.-N.; Rowan, K.; Schwinge, V.; Renzetti, L. M. N-Aroyl-L-Phenylalanine Derivatives as VCAM/VLA-4 Antagonists. *Bioorg. Med. Chem. Lett.* **2002**, *12*, 2479–2482.

53. Chiba, J.; Takayama, G.; Takashi, T.; Yokoyama, M.; Nakayama, A.; Baldwin, J. J.; McDonald, E.; Moriarty, K. J.; Sarko, C. R.; Saionz, K. W.; Swanson, R.; Hussain, Z.; Wong, A.; Machinaga, N. Synthesis, Biological Evaluation, and Pharmacokinetic Study of Prolyl-1-Piperazinylacetic Acid and Prolyl-4-Piperidinylacetic Acid Derivatives as VLA-4 Antagonists. *Bioorg. Med. Chem.* **2006**, *14*, 2725–2746.

54. Maggiora, G.; Vogt, M.; Stumpfe, D.; Bajorath, J. Molecular Similarity in Medicinal Chemistry. *J. Med. Chem.* **2014**, *57*, 3186–3204.

55. Keiser, M. J.; Setola, V.; Irwin, J. J.; Laggner, C.; Abbas, A. I.; Hufeisen, S. J.; Jensen, N. H.; Kuijer, M. B.; Matos, R. C.; Tran, T. B.; Whaley, R.; Glennon, R. A.; Hert, J.; Thomas, K. L. H.; Edwards, D. D.; Shoichet, B. K.; Roth, B. L. Predicting New Molecular Targets for Known Drugs. *Nature* **2009**, *462*, 175–181.

56. Brown, J. B.; Niijima, S.; Okuno, Y. Compound-Protein Interaction Prediction within Chemogenomics: Theoretical Concepts, Practical Usage, and Future Directions. *Mol. Inf.* **2013**, *32*, 906–921.

57. Gerard, J. P.; van Westen, J. K.; Wegner, A. P. I.; Ijzerman, A. P.; van Vlijmen, H. W. T.; Bender, A. Proteochemometric Modeling as a Tool to Design Selective Compounds and for Extrapolating to Novel Targets. *Med. Chem. Commun.* **2011**, *2*, 16–30.

58. Hopkins, A. L. Network Pharmacology: The Next Paradigm in Drug Discovery. *Nat. Chem. Biol.* **2008**, *4*, 682–690.

59. Oprea, T. I.; Tropsha, A.; Faulon, J.-L.; Rintoul, M. D. Systems Chemical Biology. *Nat. Chem. Biol.* **2007**, *3*, 447–450.

60. Wild, D. J.; Ding, Y.; Sheth, A. P.; Harland, L.; Gifford, E. M.; Lajiness, M. S. Systems Chemical Biology and the Semantic Web: What they Mean for the Future of Drug Discovery Research. *Drug Discov. Today* **2012**, *17*, 469–474.

61. Yamanishi, Y.; Araki, M.; Gutteridge, A.; Honda, W.; Kanehisa, M. Prediction of Drug-Target Interaction Networks from the Integration of Chemical and Genomic Spaces. *Bioinformatics* **2008**, *24*, i232–i240.

62. Renner, S. Past, Present and Future of Chemogenomics. *Mol. Inf.* **2013**, *32*, 877–879.

63. Kalliokoski, T.; Kramer, C.; Vulpetti, A. Quality Issues with Public Domain Chemogenomics Data. *Mol. Inf.* **2013**, *32*, 898–905.

64. https://www.ebi.ac.uk/chembl/.

65. https://pubchem.ncbi.nlm.nih.gov/.

66. https://www.broadinstitute.org/genome_bio/connectivitymap.html.

67. Ziegler, S.; Pries, V.; Hedberg, C.; Waldmann, H. Target Identification for Small Bioactive Molecules: Finding the Needle in the Haystack. *Angew. Chem. Int. Ed.* **2013**, *52*, 2744–2792.

68. Giaever, G.; Shoemaker, D. D.; Jones, T. W.; Liang, H.; Winzeler, E. A.; Astromoff, A.; Davis, R. W. Genomic Profiling of Drug Sensitivities Via Induced Haploinsufficiency. *Nat. Genet.* **1999**, *21*, 278–283.

69. Giaever, G.; Flaherty, P.; Kumm, J.; Proctor, M.; Nislow, C.; Jaramillo, D. F.; Chu, A. M.; Jordan, M. I.; Arkin, A. P.; Davis, R. W. Chemogenomic Profiling: Identifying the Functional Interactions of Small Molecules in Yeast. *Proc. Natl. Acad. Sci. USA* **2004**, *101*, 793–798.

70. Hillenmeyer, M. E.; Fung, E.; Wildenhain, J.; Pierce, S. E.; Hoon, S.; Lee, W.; Proctor, M.; St Onge, R. P.; Tyers, M.; Koller, D.; Altman, R. B.; Davis, R. W.; Nislow, C.; Giaever, G. The Chemical Genomic Portrait of Yeast: Uncovering a Phenotype for All Genes. *Science* **2008**, *320*, 362–365.

71. Wilmes, A.; Hanna, R.; Heathcott, R. W.; Northcote, P. T.; Atkinson, P. H.; Bellows, D. S.; Miller, J. H. Chemical Genetic Profiling of the Microtubule-Targeting Agent Peloruside A in Budding Yeast *Saccharomyces cerevisiae*. *Gene* **2012**, *497*, 140–146.

72. Butcher, R. A.; Schreiber, S. L. A Microarray-Based Protocol for Monitoring the Growth of Yeast Overexpression Strains. *Nat. Protoc.* **2006**, *1*, 569–576.

73. Hei Ho, C.; Magtanong, L.; Barker, S. L.; Gresham, D.; Nishimura, S.; Natarajan, P.; Koh, J. L. Y.; Porter, J.; Gray, C. A.; Andersen, R. J.; Giaever, G.; Nislow, C.; Andrews, B.; Botstein, D.; Graham, T. R.; Yoshida, M.; Boone, C. A Molecular Barcoded Yeast ORF Library Enables Mode-of-Action Analysis of Bioactive Compounds. *Nat. Biotech.* **2009**, *27*, 369–377.

74. Hoon, S.; Smith, A. M.; Wallace, I. M.; Suresh, S.; Miranda, M.; Fung, E.; Proctor, M.; Shokat, K. M.; Zhang, C.; Davis, R. W.; Giaever, G.; St Onge, R. P.; Nislow, C. An Integrated Platform of Genomic Assays Reveals Small-Molecule Bioactivities. *Nat. Chem. Biol.* **2008**, *4*, 498–506.

75. Castoreno, A. B.; Smurnyy, Y.; Torres, A. D.; Vokes, M. S.; Jones, T. R.; Carpenter, A. E.; Eggert, U. S. Small Molecules Discovered in a Pathway Screen Target the Rho Pathway in Cytokinesis. *Nat. Chem. Biol.* **2010**, *6*, 457–463.

76. Wang, J.; Zhou, X.; Bradley, P. L.; Chang, S.-F.; Perrimon, N.; Wong, S. T. C. Cellular Phenotype Recognition for High-Content RNA Interference Genome-Wide Screening. *J. Biomol. Screen.* **2008**, *13*, 29–39.

77. Matheny, C. J.; Wei, M. C.; Bassik, M. C.; Donnelly, A. J.; Kampmann, M.; Iwasaki, M.; Piloto, O.; Solow-Cordero, D. E.; Bouley, D. M.; Rau, R.; Brown, P.; McManus, M. T.; Weissman, J. S.; Cleary, M. L. Next-Generation NAMPT Inhibitors Identified by Sequential High-Throughput Phenotypic Chemical and Functional Genomic Screens. *Chem. Biol.* **2013**, *20*, 1352–1363.

78. Friedman, A.; Perrimon, N. Genome-Wide High-Throughput Screens in Functional Genomics. *Curr. Opin. Genet. Dev.* **2004**, *14*, 470–476.

79. Echeverri, C. J.; Perrimon, N. High-Throughput RNAi Screening in Cultured Cells: A User's Guide. *Nat. Rev. Genet.* **2006**, *7*, 373–384.

80. Hung-Chieh Chou, D.; Vetere, A.; Choudhary, A.; Scully, S. S.; Schenone, M.; Tang, A.; Gomez, R.; Burns, S. M.; Lundh, M.; Vital, T.; Comer, E.; Faloon, P. W.; Dančík, V.; Ciarlo, C.; Paulk, J.; Dai, M.; Reddy, C.; Sun, H.; Young, M.; Donato, N.; Jaffe, J.; Clemons, P. A.; Palmer, M.; Carr, S. A.; Schreiber, S. L.; Wagner, B. K. Kinase-Independent Small-Molecule Inhibition of JAK-STAT Signaling. *J. Am. Chem. Soc.* **2015**, *137*, 7929–7934.

81. Yoo, Y.-H.; Yun, J. H.; Yoon, C. N.; Lee, J.-S. Chemical Proteomic Identification of T-Plastin as a Novel Host Cell Response Factor in HCV Infection. *Sci. Rep.* **2015**, *5*, 9773.

82. Prosser, G. A.; Larrouy-Maumus, G.; de Carvalho, L. P. S. Metabolomic Strategies for the Identification of New Enzyme Functions and Metabolic Pathways. *EMBO Rep.* **2014**, *15*, 657–669.

83. Perlman, Z. E.; Slack, M. D.; Feng, Y.; Mitchison, T. J.; Wu, L. F.; Altschuler, S. J. Multidimensional Drug Profiling by Automated Microscopy. *Science* **2004**, *306*, 1194–1198.

84. Abassi, Y. A.; Xi, B.; Zhang, W.; Ye, P.; Kirstein, S. L.; Gaylord, M. R.; Feinstein, S. C.; Wang, X.; Xu, X. Kinetic Cell-Based Morphological Screening: Prediction of Mechanism of Compound Action and Off-Target Effects. *Chem. Biol.* **2009**, *16*, 712–723.

85. Muroi, M.; Kazami, S.; Noda, K.; Kondo, H.; Takayama, H.; Kawatani, M.; Usui, T.; Osada, H. Application of Proteomic Profiling Based on 2D-DIGE for Classification of Compounds According to the Mechanism of Action. *Chem. Biol.* **2010**, *17*, 460–470.

86. Lomenick, B.; Hao, R.; Jonai, N.; Chin, R. M.; Aghajan, R.; Warburton, S.; Wang, J.; Wu, R. P.; Gomez, F.; Loo, J. A.; Wohlschlegel, J. A.; Vondriska, T. M.; Pelletier, J.; Herschman, H. R.; Clardy, J.; Clarke, C. F.; Huang, J. Target Identification Using Drug Affinity Responsive Target Stability (DARTS). *Proc. Natl. Acad. Sci. USA* **2009**, *106*, 21984–21989.

87. Lomenick, B.; Olsen, R. W.; Huang, J. Identification of Direct Protein Targets of Small Molecules. *ACS Chem. Biol.* **2011**, *6*, 34–46.

88. Aghajan, M.; Jonai, N.; Flick, K.; Fu, F.; Luo, M.; Cai, X.; Ouni, I.; Pierce, N.; Tang, X.; Lomenick, B.; Damoiseaux, R.; Hao, R.; del Moral, P. M.; Verma, R.; Li, Y.; Li, C.; Houk, K. N.; Jung, M. E.; Zheng, N.; Huang, L.; Deshaies, R. J.; Kaiser, P.; Huang, J. Chemical Genetics Screen for Enhancers of Rapamycin Identifies a Specific Inhibitor of an SCF Family E3 Ubiquitin Ligase. *Nat. Biotech.* **2010**, *28*, 738–742.

89. West, G. M.; Tucker, C. L.; Xu, T.; Park, S.-K.; Han, X.; Yates, J. R., III; Fitzgerald, M. C. Quantitative Proteomics Approach for Identifying Protein-Drug Interactions in Complex Mixtures Using Protein Stability Measurements. *Proc. Natl. Acad. Sci. USA* **2010**, *107*, 9078–9082.

90. Chan, J. N.; Vuckovic, D.; Sleno, L.; Olsen, J. B.; Pogoutse, O.; Havugimana, P.; Hewel, J. A.; Bajaj, N.; Wang, Y.; Musteata, M. F.; Nislow, C.; Emili, A. Target Identification by Chromatographic Co-Elution: Monitoring of Drug-Protein Interactions without Immobilization or Chemical Derivatization. *Mol. Cell Proteomics* **2012**, *11*, M111.016642.

91. Salcius, M.; Bauer, A. J.; Hao, Q.; Li, S.; Tutter, A.; Raphael, J.; Jahnke, W.; Rondeau, J.-M.; Bourgier, E.; Tallarico, J.; Michaud, G. A. SEC-TID: A Label-Free Method for Small Molecule Target Identification. *J. Biomol. Scr.* **2014**, *19*, 917–927.

92. Sun, D.; He, Q.-Y. Chemical Proteomics to Identify Molecular Targets of Small Compounds. *Curr. Mol. Med.* **2013**, *13*, 1175–1191.

93. Sato, S.-I.; Murata, A.; Shirakawa, T.; Uesugi, M. Biochemical Target Isolation For Novices: Affinity-Based Strategies. *Chem. Biol.* **2010**, *17*, 616–623.

94. Chou, D. H.; Duvall, J. R.; Gerard, B.; Liu, H.; Pandya, B. A.; Suh, B. C.; Forbeck, E. M.; Faloon, P.; Wagner, B. K.; Marcaurelle, L. A. Synthesis of a Novel Suppressor of β-Cell Apoptosis Via Diversity-Oriented Synthesis. *ACS Med. Chem. Lett.* **2011**, *2*, 698–702.

95. Chou, D. H.; Bodycombe, N. E.; Carrinski, H. A.; Lewis, T. A.; Clemons, P. A.; Schreiber, S. L.; Wagner, B. K. Small-Molecule Suppressors of Cytokine-Induced β-Cell Apoptosis. *ACS Chem. Biol.* **2010**, *5*, 729–734.

96. Faloon, P. W.; Chou, D. H. C.; Forbeck, E. M.; Walpita, D.; Morgan, B.; Buhrlage, S.; Ting, A.; Perez, J.; MacPherson, L. J.; Duvall, J. R.; Dandapani, S.; Marcaurelle, L. A.; Munoz, B.; Palmer, M.; Foley, M.; Wagner, B.; Schreiber, S. L. Identification of Small Molecule Inhibitors that Suppress Cytokine-Induced Apoptosis in Human Pancreatic Islet Cells. In *Probe Reports from the NIH Molecular Libraries Program*; National Center for Biotechnology Information: Bethesda, MD, 2010.

97. Shiyama, T.; Furuya, M.; Yamazaki, A.; Terada, T.; Tanaka, A. Design and Synthesis of Novel Hydrophilic Spacers for the Reduction of Nonspecific Binding Proteins on Affinity Resins. *Bioorg. Med. Chem.* **2004**, *12*, 2831–2841.

98. Sato, S.; Kwon, Y.; Kamisuki, S.; Srivastava, N.; Mao, Q.; Kawazoe, Y.; Uesugi, M. Polyproline-Rod Approach to Isolating Protein Targets of Bioactive Small Molecules: Isolation of a New Target of Indomethacin. *J. Am. Chem. Soc.* **2007**, *129*, 873–880.

99.   Sakamoto, S.; Hatakeyama, M.; Ito, T.; Handa, H. Tools and Methodologies Capable of Isolating and Identifying a Target Molecule for a Bioactive Compound. *Bioorg. Med. Chem.* **2012**, *20*, 1990–2001.

100.   Shimizu, N.; Sugimoto, K.; Tang, J.; Nishi, T.; Sato, I.; Hiramoto, M.; Aizawa, S.; Hatakeyama, M.; Ohba, R.; Hatori, H.; Yoshikawa, T.; Suzuki, F.; Oomori, A.; Tanaka, H.; Kawaguchi, H.; Watanabe, H.; Handa, H. High-Performance Affinity Beads for Identifying Drug Receptors. *Nat. Biotechnol.* **2000**, *18*, 877–881.

101.   Botubol Ares, J. M.; Duran-Pena, M. J.; Hernandez-Galan, R.; Collado, I. G. Chemical Genetics Strategies for Identification of Molecular Targets. *Phytochem. Rev.* **2013**, *12*, 895–914.

102.   Kanoh, N.; Honda, K.; Simizu, S.; Muroi, M.; Osada, H. Photo-Cross-Linked Small-Molecule Affinity Matrix for Facilitating Forward and Reverse Chemical Genetics. *Angew. Chem. Int. Ed.* **2005**, *44*, 3559–3562.

103.   Kanoh, N.; Takayama, H.; Honda, K.; Moriya, T.; Teruya, T.; Simizu, S.; Osada, H.; Iwabuchi, Y. Cleavable Linker for Photo-Cross-Linked Small Molecule Affinity Matrix. *Biocon. Chem.* **2010**, *21*, 182–186.

104.   Kawatani, M.; Okumura, H.; Honda, K.; Kanoh, N.; Muroi, M.; Dohmae, N.; Takami, M.; Kitagawa, M.; Futamura, Y.; Imoto, M.; Osada, H. The Identification of an Osteoclastogenesis Inhibitor through the Inhibition of Glyoxalase I. *Proc. Natl. Acad. Sci. USA* **2008**, *105*, 11691–11696.

105.   Kondoh, Y.; Osada, H. High-Throughput Screening Identifies Small Molecule Inhibitors of Molecular Chaperones. *Curr. Pharm. Des.* **2013**, *19*, 473–492.

106.   Kotake, Y.; Sagane, K.; Owa, T.; Mimori-Kiyosue, Y.; Shimizu, H.; Uesugi, M.; Ishihama, Y.; Iwata, M.; Mizui, Y. Splicing Factor SF3b as a Target of the Antitumor Natural Product Pladienolide. *Nat. Chem. Biol.* **2007**, *3*, 570–575.

107.   Yamamoto, K.; Yamazaki, A.; Takeuchi, M.; Tanaka, A. A. A Versatile Method of Identifying Specific Binding Proteins on Affinity Resins. *Anal. Biochem.* **2006**, *352*, 15–23.

108.   Harding, M. W.; Galat, A.; Uehling, D. E.; Schreiber, S. L. A Receptor for the Immunosuppressant FK506 Is a *Cis-Trans* Peptidyl-Prolyl Isomerase. *Nature* **1989**, *341*, 758–760.

109.   Verhelst, S. H.; Fonovic, M.; Bogyo, M. A Mild Chemically Cleavable Linker System for Functional Proteomic Applications. *Angew. Chem. Int. Ed.* **2007**, *46*, 1284–1286.

110.   Koopmans, T.; Dekker, F. J.; Martin, N. I. A Photocleavable Affinity Tag for the Enrichment of Alkyne-Modified Biomolecules. *RSC Adv.* **2012**, *2*, 2244–2246.

111.   Sato, S.; Murata, A.; Orihara, T.; Shirakawa, T.; Suenaga, K.; Kigoshi, H.; Uesugi, M. Marine Natural Product Aurilide Activates the OPA1-Mediated Apoptosis by Binding to Prohibitin. *Chem. Biol.* **2011**, *18*, 131–139.

112.   Fleischer, T. C.; Murphy, B. R.; Flick, J. S.; Terry-Lorenzo, R. T.; Gao, Z. H.; Davis, T.; McKinnon, R.; Ostanin, K.; Willardsen, J. A.; Boniface, J. J. Chemical Proteomics Identifies Nampt as the Target of CB30865, an Orphan Cytotoxic Compound. *Chem. Biol.* **2010**, *17*, 659–664.

113.   Karpievitch, Y. V.; Polpitiya, A. D.; Anderson, G. A.; Smith, R. D.; Dabney, A. R. Liquid Chromatography Mass Spectrometry-Based Proteomics: Biological and Technological Aspects. *Ann. Appl. Stat.* **2010**, *4*(4), 1797–1823.

114.   Han, X.; Aslanian, A.; Yates, J. R., III. Mass Spectrometry for Proteomics. *Curr. Opin. Chem. Biol.* **2008**, *12*, 483–490.

115.   Ong, S.-E.; Blagoev, B.; Kratchmarova, I.; Kristensen, D. B.; Steen, H.; Pandey, A.; Mann, M. Stable Isotope Labeling by Amino Acids in Cell Culture, SILAC, as a Simple and Accurate Approach to Expression Proteomics. *Mol. Cell. Proteomics* **2002**, *1*, 376–386.

116. Ong, S.-E.; Schenone, M.; Margolin, A. A.; Li, X.; Do, K.; Doud, M. K.; Mani, D. R.; Kuai, L.; Wang, X.; Wood, J. L.; Tolliday, N. J.; Koehler, A. N.; Marcaurelle, L. A.; Golub, T. R.; Gould, R. J.; Schreiber, S. L.; Carr, S. A. Identifying the Proteins to which Small-Molecule Probes and Drugs Bind in Cells. *Proc. Natl. Acad. Sci. USA* **2009**, *106*, 4617–4622.

117. Gygi, S. P.; Rist, B.; Gerber, S. A.; Turecek, F.; Gelb, M. H.; Aebersold, R. Quantitative Analysis of Complex Protein Mixtures Using Isotope-Coded Affinity Tags. *Nat. Biotechnol.* **1999**, *17*, 994–999.

118. Yi, E. C.; Li, X. J.; Cooke, K.; Lee, H.; Raught, B.; Page, A.; Aneliunas, V.; Hieter, P.; Goodlett, D. R.; Aebersold, R. Increased Quantitative Proteome Coverage with (13)C/(12)C-Based, Acid-Cleavable Isotope-Coded Affinity Tag Reagent and Modified Data Acquisition Scheme. *Proteomics* **2005**, *5*, 380–387.

119. Bantscheff, M.; Eberhard, D.; Abraham, Y.; Bastuck, S.; Boesche, M.; Hobson, S.; Mathieson, T.; Perrin, J.; Raida, M.; Rau, C.; Reader, V.; Sweetman, G.; Bauer, A.; Bouwmeester, T.; Hopf, C.; Kruse, U.; Neubauer, G.; Ramsden, N.; Rick, J.; Kuster, B.; Drewes, G. Quantitative Chemical Proteomics Reveals Mechanisms of Action of Clinical ABL Kinase Inhibitors. *Nat. Biotechnol.* **2007**, *25*, 1035–1044.

120. Kawatami, M.; Osada, H. Affinity-Based Target Identification for Bioactive Small Molecules. *Med. Chem. Commun.* **2014**, *5*, 277–287.

121. Liu, J.; Farmer, J. D.; Lane, W. S.; Friedman, J.; Weissman, I.; Schreiber, S. L. Calcineurin Is a Common Target of Cyclophilin-Cyclosporin a and FKBP-FK506 Complexes. *Cell* **1991**, *66*, 807–815.

122. Trippier, P. C. Synthetic Strategies for the Biotinylation of Bioactive Small Molecules. *ChemMedChem* **2013**, *8*, 190–203.

123. Wirth, T.; Schmuck, K.; Tietze, L. F.; Sieber, S. A. Novel Mode of Action: Duocarmycin Analogs Target Aldehyde dehydrogenase1 in Lung Cancer Cells. *Angew. Chem. Int. Ed.* **2012**, *51*, 2874–2877.

124. Raghavan, A. S.; Hang, A. C. Seeing Small Molecules in Action with Bioorthogonal Chemistry. *Drug Disc. Today* **2009**, *14*, 178–184.

125. Sheih, P.; Bertozzi, C. R. Design Strategies for Biorthogonal Smart Probes. *Org. Biomol. Chem.* **2014**, *12*, 9307–9320.

126. Hein, J. E.; Fokin, V. V. Copper-Catalyzed Azide–Alkyne Cycloaddition (CuAAC) and beyond: New Reactivity of Copper(I) Acetylides. *Chem. Soc. Rev.* **2010**, *39*, 1302–1315.

127. Cheng, X.; Li, L.; Uttamchandani, M.; Yao, S. Q. In Situ Proteome Profiling of C75, a Covalent Bioactive Compound with Potential Anticancer Activities. *Org. Lett.* **2014**, *16*, 1414–1417.

128. Schilling, C. I.; Jung, N.; Biskup, M.; Schepers, U.; Brase, S. Bioconjugation Via Azide–Staudinger Ligation: An Overview. *Chem. Soc. Rev.* **2011**, *40*, 4840–4871.

129. Stubbs, K. A.; Scaffidi, A.; Debowski, A. W.; Mark, B. L.; Stick, R. V.; Vocadlo, D. J. Synthesis and Use of Mechanism-Based Protein-Profiling Probes for Retaining Beta-D-Glucosaminidases Facilitate Identification of Pseudomonas Aeruginosa NagZ. *J. Am. Chem. Soc.* **2008**, *130*, 327–335.

130. Yang, K. S.; Budin, G.; Reiner, T.; Vinegoni, C.; Weissleder, R. Bioorthogonal Imaging of Aurora Kinase a in Live Cells. *Angew. Chem. Int. Ed.* **2012**, *51*, 6598–6603.

131. Su, Y.; Pan, S.; Li, Z.; Li, L.; Wu, X.; Hao, P.; Sze, S. K.; Yao, S. Q. Multiplex Imaging and Cellular Target Identification of Kinase Inhibitors Via an Affinity-Based Proteome Profiling Approach. *Sci. Rep.* **2015**, *5*, 7724.

132. Niphakis, M. J.; Cravatt, B. F. Enzyme Inhibitor Discovery by Activity-Based Protein Profiling. *Annu. Rev. Biochem.* **2014**, *83*, 341–377.

133. Yang, P.; Liu, K. Activity-Based Protein Profiling: Recent Advances in Probe Development and Applications. *ChemBioChem* **2015**, *16*, 712–724.

134. Liu, N.; Hoogendoorn, S.; van de Kar, B.; Kaptein, A.; Barf, T.; Driessen, C.; Filippov, D. V.; van der Marel, G. A.; van der Stelt, M.; Overkleeft, H. S. Direct and two-Step Bioorthogonal Probes for Bruton's Tyrosine Kinase Based on Ibrutinib: A Comparative Study. *Org. Biomed. Chem.* **2015**, *13*, 5147–5157.

135. Sumranjit, J.; Chung, S. J. Recent Advances in Target Characterization and Identification by Photoaffinity Probes. *Molecules* **2013**, *18*, 10425–10451.

136. Dorman, G.; Prestwich, G. D. Benzophenone Photophores in Biochemistry. *Biochemistry* **1994**, *33*, 5661–5673.

137. Nielsen, P. E.; Buchardt, O. Aryl Azides as Photoaffinity Labels. A Photochemical Study of some 4-Substituted Aryl Azides. *Photochem. Photobiol.* **1982**, *35*, 317–323.

138. Das, J. Aliphatic Diazirines as Photoaffinity Probes for Proteins: Recent Developments. *Chem. Rev.* **2011**, *111*, 4405–4444.

139. Liu, K.; Shi, H.; Xiao, H.; Chong, A. G.; Bi, X.; Chang, Y.-T.; Wan, K. S. W.; Yada, R. Y.; Yao, S. Q. *Angew. Chem. Int. Ed.* **2009**, *48*, 8293–8297.

140. Rotili, D.; Altun, M.; Kawamura, A.; Wolf, A.; Fischer, R.; Leung, I. K. H.; Mackeen, M. M.; Tian, J.-m.; Ratcliffe, P. J.; Mai, A.; Kessler, B. M.; Schofield, C. *J. Chem. Biol.* **2011**, *18*, 642–654.

141. MacKinnon, A. L.; Garrison, J. L.; Hegde, R. S.; Taunton, J. *J. Am. Chem. Soc.* **2007**, *129*, 14560–14561.

142. Fischer, J. J.; Graebner Baessler, O. Y.; Dalhoff, C.; Michaelis, S.; Schrey, A. K.; Ungewiss, J.; Andrich, K.; Jeske, D.; Kroll, F.; Glinski, M.; Sefkow, M.; Dreger, M.; Koester, H. Comprehensive Identification of Staurosporine-Binding Kinases in the Hepatocyte Cell Line HepG2 Using Capture Compound Mass Spectrometry (CCMS). *J. Proteome Res.* **2010**, *9*, 806–817.

143. Fischer, J. J.; Michaelis, S.; Schrey, A. K.; Graebner Baessler, O. Y.; Glinski, M.; Dreger, M.; Kroll, F.; Koester, H. Capture Compound Mass Spectrometry Sheds Light on the Molecular Mechanisms of Liver Toxicity of Two Parkinson Drugs. *Toxicol. Sci.* **2010**, *113*, 243–253.

144. Kambe, T.; Correia, B. E.; Niphakis, M. J.; Cravatt, B. F. Mapping the Protein Interaction Landscape for Fully Functionalized Small-Molecule Probes in Human Cells. *J. Am. Chem. Soc.* **2014**, *136*, 10777–10782.

145. Clackson, T.; Wells, J. A. In Vitro Selection from Protein and Peptide Libraries. *Trends Biotechnol.* **1994**, *12*, 173–184.

146. Piggott, A. M.; Karuso, P. Rapid Identification of a Protein Binding Partner for the Marine Natural Product Kahalalide F by Using Reverse Chemical Proteomics. *ChemBioChem* **2008**, *9*, 524–530.

147. Licitra, E. J.; Liu, J. O. A Three-Hybrid System for Detecting Small Ligand-Protein Receptor Interactions. *Proc. Natl. Acad. Sci. USA* **1996**, *93*, 12817–12821.

148. Becker, F.; Murthi, K.; Smith, C.; Come, J.; Costa-Roldàn, N.; Kaufmann, C.; Hanke, U.; Degenhart, C.; Baumann, S.; Wallner, W.; Huber, A.; Dedier, S.; Dill, S.; Kinsman, D.; Hediger, M.; Bockovich, N.; Meier-Ewert, S.; Kluge, A. F.; Kley, N. *Chem. Biol.* **2004**, *11*, 211–223.

149. Moser, S.; Johnsson, K. Yeast three-Hybrid Screening for Identifying Anti-Tuberculosis Drug Targets. *ChemBioChem* **2013**, *14*, 2239–2242.

150. Redon, C.; Pilch, D.; Rogakou, E.; Sedelnikova, O.; Newrock, K.; Bonner, W. Histone H2A Variants H2AX and H2AZ. *Curr. Opin. Genet. Dev.* **2002**, *12*, 162–169.

151. Kornberg, R. D. Chromatin Structure: A Repeating Unit of Histones and DNA. *Science* **1974**, *184*, 868–871.

152. MacAlpine, D. M.; Almouzni, G. Chromatin and DNA Replication. *Cold Spring Harb. Perspect. Biol.* **2013**, *5*, a010207.

153. Luger, K.; Mader, A. W.; Richmond, R. K.; Sargent, D. F.; Richmond, T. J. Crystal Structure of the Nucleosome Core Particle at 2.8 Å Resolution. *Nature* **1997**, *389*, 251–260.

154. Maeshima, K.; Imai, R.; Tamura, S.; Nozaki, T. Chromatin as Dynamic 10-Nm Fibers. *Chromosoma* **2014**, *123*, 225–237.

155. Luger, K.; Richmond, T. J. The Histone Tails of the Nucleosome. *Curr. Opin. Genet. Dev.* **1998**, *8*, 140–146.

156. Luger, K.; Dechassa, M. L.; Tremethick, D. J. New Insights into Nucleosome and Chromatin Structure: An Ordered State or a Disordered Affair? *Nat. Rev. Mol. Cell Biol.* **2012**, *13*, 436–447.

157. Li, G.; Zhu, P. Structure and Organization of Chromatin Fiber in the Nucleus. *FEBS Lett.* **2015**, *589*, 2893–2904.

158. Cooper, G. M. Chapter 5.3: Chromosomes and Chromatin. In *The Cell: A Molecular Approach*; 6th ed.; Sinauer Associates: Sunderland, MA, USA, 2013; pp 171–180.

159. Rothbart, S. B.; Strahl, B. D. Interpreting the Language of Histone and DNA Modifications. *Biochim. Biophys. Acta* **2014**, *1839*, 627–643.

160. Kebede, A. F.; Schneider, R.; Daujat, S. Novel Types and Sites of Histone Modifications Emerge as Players in the Transcriptional Regulation Contest. *FEBS Lett.* **2015**, *282*, 1658–1674.

161. Allfrey, V. G.; Faulkner, R.; Mirsky, A. E. Acetylation and Methylation of Histones and their Possible Role in the Regulation of RNA Synthesis. *Proc. Natl. Acad. Sci. USA* **1964**, *51*, 786–794.

162. Gutierrez, R. M.; Hnilica, L. S. Tissue Specificity of Histone Phosphorylation. *Science* **1967**, *157*, 1324–1325.

163. Arnaudo, A. M.; Garcia, B. A. Proteomic Characterization of Novel Histone Posttranslational Modifications. *Epigen. Chrom.* **2013**, *6*, 24.

164. Hong, L.; Schroth, G. P.; Matthews, H. R.; Yau, P.; Bradbury, E. M. Studies of the DNA Binding Properties of Histone H4 Amino Terminus. Thermal Denaturation Studies Reveal that Acetylation Markedly Reduces the Binding Constant of the H4 "Tail" to DNA. *J. Biol. Chem.* **1993**, *268*, 305–314.

165. Zhang, Y.; Reinberg, D. Transcription Regulation by Histone Methylation: Interplay between Different Covalent Modifications of the Core Histone Tails. *Genes Dev.* **2001**, *15*, 2343–2360.

166. Krogan, N. J.; Dover, J.; Wood, A.; Schneider, J.; Heidt, J.; Boateng, M. A.; Dean, K.; Ryan, O. W.; Golshani, A.; Johnston, M.; Greenblatt, J. F.; Shilatifard, A. The Paf1 Complex Is Required for Histone H3 Methylation by COMPASS and Dot1p: Linking Transcriptional Elongation to Histone Methylation. *Mol. Cell* **2003**, *11*, 721–729.

167. Jorgensen, S.; Schotta, G.; Sorensen, C. S. Histone H4 Lysine 20 Methylation: Key Player in Epigenetic Regulation of Genomic Integrity. *Nucl. Acids Res.* **2013**, *41*, 2797–2806.

168. Zhou, J.; Wang, X.; He, K.; Charron, J. B.; Elling, A. A.; Deng, X. W. Genome-Wide Profiling of Histone H3 Lysine 9 Acetylation and Dimethylation in Arabidopsis Reveals Correlation Between Multiple Histone Marks and Gene Expression. *Plant Mol. Biol.* **2010**, *72*, 585–595.

169. Young, M. D.; Willson, T. A.; Wakefield, M. J.; Trounson, E.; Hilton, D. J.; Blewitt, M. E.; Oshlack, A.; Majewski, I. J. ChIP-seq Analysis Reveals Distinct H3K27me3 Profiles That Correlate with Transcriptional Activity. *Nucl. Acids Res.* **2011**, *39*, 7415–7427.

170. Lienert, F.; Mohn, F.; Tiwari, V. K.; Baubec, T.; Roloff, T. C.; Gaidatzis, D.; Stadler, M. B.; Schübeler, D. Genomic Prevalence of Heterochromatic H3K9me2 and Transcription Do Not Discriminate Pluripotent from Terminally Differentiated Cells. *PLoS Genet.* **2011**, *7*, e1002090.

171. Johansen, K. M.; Johansen, J. Regulation of Chromatin Structure by Histone H3S10 Phosphorylation. *Chromosome Res.* **2006**, *14*, 393–404.

172. Chapman-Rothe, N.; Curry, E.; Zeller, C.; Liber, D.; Stronach, E.; Gabra, H.; Ghaem-Maghami, S.; Brown, R. Chromatin H3K27me3/H3K4me3 Histone Marks Define Gene Sets in High-Grade Serous Ovarian Cancer that Distinguish Malignant, Tumour-Sustaining and Chemo-Resistant Ovarian Tumour Cells. *Oncogene* **2013**, *32*, 4586–4592.

173. Wang, Z.; Zang, C.; Rosenfeld, J. A.; Schones, D. E.; Barski, A.; Cuddapah, S.; Cui, K.; Roh, T.-Y.; Peng, W.; Zhang, M. Q.; Zhao, K. Combinatorial Patterns of Histone Acetylations and Methylations in the Human Genome. *Nat. Genet.* **2008**, (7), 897–903.

174. Swygert, S. G.; Peterson, C. L. Chromatin Dynamics: Interplay Between Remodeling Enzymes and Histone Modifications. *Biochim. Biophys. Acta* **2014**, *1839*, 728–736.

175. Strahl, B. D.; Allis, C. D. The Language of Covalent Histone Modifications. *Nature* **2000**, *403*, 41–45.

176. Trievel, R. C. Structure and Function of Histone Methyltransferases. *Crit. Rev. Eukaryot. Gene Expr.* **2004**, *14*, 147–169.

177. Roth, S. Y.; Denu, J. M.; Allis, C. D. Histone Acetyltransferases. *Annu. Rev. Biochem.* **2001**, *70*, 81–120.

178. Baek, S. H. When Signaling Kinases Meet Histones and Histone Modifiers in the Nucleus. *Mol. Cell* **2011**, *42*, 274–284.

179. Rotili, D.; Mai, A. Targeting Histone Demethylases: A New Avenue for the Fight Against Cancer. *Genes Cancer* **2011**, *2*, 663–679.

180. Delcuve, G. P.; Khan, D. H.; Davie, J. R. Roles of Histone Deacetylases in Epigenetic Regulation: Emerging Paradigms from Studies with Inhibitors. *Clin. Epigenetics* **2012**, *4*, 5.

181. Rossetto, D.; Avvakumov, N.; Côté, J. Histone Phosphorylation: A Chromatin Modification Involved in Diverse Nuclear Events. *Epigenetics* **2012**, *7*, 1098–1108.

182. Musselman, C. A.; Lalonde, M.-E.; Côté, J.; Kutateladze, T. G. Perceiving the Epigenetic Landscape through Histone Readers. *Nat. Struct. Mol. Biol.* **2012**, *19*, 1218–1227.

183. Sanchez, R.; Meslamani, J.; Zhou, M.-M. The Bromodomain: From Epigenome Reader to Druggable Target. *Biochim. Biophys. Acta* **2014**, *1839*, 676–685.

184. Li, Y.; Li, H. Many Keys to Push: Diversifying the 'Readership' of Plant Homeodomain Fingers. *Acta Biochim. Biophys. Sinica* **2012**, *44*, 28–39.

185. Jones, D. O.; Cowell, I. G.; Singh, P. B. Mammalian Chromodomain Proteins: Their Role in Genome Organisation and Expression. *Bioessays* **2000**, *22*, 124–137.

186. Ying, M.; Chen, D. Tudor Domain-Containing Proteins of Drosophila melanogaster. *Dev. Growth Differ.* **2012**, *54*, 32–43.

187. Winter, S.; Simboeck, E.; Fischle, W.; Zupkovitz, G.; Dohnal, I.; Mechtler, K.; Ammerer, G.; Seiser, C. 14-3-3 Proteins Recognize a Histone Code at Histone H3 and are Required for Transcriptional Activation. *EMBO J.* **2008**, *27*, 88–99.

188. Geiman, T. M.; Robertson, K. D. Chromatin Remodeling, Histone Modifications, and DNA Methylation-How Does It All Fit Together? *J. Cell Biochem.* **2002**, *87*, 117–125.

189. Talbert, P. B.; Henikoff, S. Histone Variants—Ancient Wrap Artists of the Epigenome. *Nat. Rev. Mol. Cell Biol.* **2010**, *11*, 264–275.

190. Stein, G. S.; van Wijnen, A. J.; Imbalzano, A. N.; Montecino, M.; Zaidi, S. K.; Lian, J. B.; Nickerson, J. A.; Stein, J. L. Architectural genetic and epigenetic control of regulatory networks: compartmentalizing machinery for transcription and chromatin remodeling in nuclear microenvironments. *Crit. Rev. Eukaryot. Gene Expr.* **2010**, *20*, 149–155.

191. Barth, T. K.; Imhof, A. Fast Signals and Slow Marks: The Dynamics of Histone Modifications. *Trends Biochem. Sci.* **2010**, *35*, 618–626.

192. Mottamal, M.; Zheng, S.; Huang, T. L.; Wang, G. Histone Deacetylase Inhibitors in Clinical Studies as Templates for New Anticancer Agents. *Molecules* **2015**, *20*, 3898–3941.

193. Wang, C.-Y.; Filippakopoulos, P. BETting the Odds: BET in Disease. *Trends Biochem. Sci.* **2015**, *40*, 468–479.

194. Maes, T.; Carceller, E.; Salas, J.; Ortega, A.; Buesa, C. Advances in the Development of Histone Lysine Demethylase Inhibitors. *Curr. Opin. Pharmacol.* **2015**, *23*, 52–60.

195. Zagni, C.; Chiacchio, U.; Rescifina, A. Histone Methyltransferase Inhibitors: Novel Epigenetic Agents for Cancer Treatment. *Curr. Med. Chem.* **2013**, *20*, 167–185.

196. Furdas, S. D.; Kannan, S.; Sippl, W.; Jung, M. Small Molecule Inhibitors of Histone Acetyltransferases as Epigenetic Tools and Drug Candidates. *Arch. Pharm.* **2012**, *345*, 7–21.

197. Sperandio, O.; Reynes, C. H.; Camproux, A. C.; Villoutreix, B. O. Rationalizing the Chemical Space of Protein-Protein Interaction Inhibitors. *Drug Discov. Today* **2009**, *15*, 220–229.

198. Jang, M. K.; Mochizuki, K.; Zhou, M.; Jeong, H.-S.; Brady, J. N.; Ozato, K. The Bromodomain Protein Brd4 Is a Positive Regulatory Component of P-TEFb and Stimulates RNA Polymerase II-Dependent Transcription. *Mol. Cell* **2005**, *19*, 523–534.

199. LeRoy, G.; Rickards, B.; Flint, S. J. The Double Bromodomain Proteins Brd2 and Brd3 Couple Histone Acetylation to Transcription. *Mol. Cell* **2008**, *30*, 51–60.

200. Bisgrove, D. A.; Mahmoudi, T.; Henklein, P.; Verdin, E. Conserved P-TEFb-Interacting Domain of BRD4 Inhibits HIV Transcription. *Proc. Natl Acad. Sci. USA* **2007**, *104*, 13690–13695.

201. French, C. A.; Miyoshi, I.; Kubonishi, I.; Grier, H. E.; Perez-Atayde, A. R.; Fletcher, J. A. BRD4-NUT Fusion Oncogene: A Novel Mechanism in Aggressive Carcinoma. *Cancer Res.* **2003**, *63*, 304–307.

202. Nicodeme, E.; Jeffrey, K. L.; Schaefer, U.; Beinke, S.; Dewell, S.; Chung, C.-W.; Chandwani, R.; Marazzi, I.; Wilson, P.; Coste, H.; White, J.; Kirilovsky, J.; Rice, C. M.; Lora, J. M.; Prinjha, R. K.; Lee, K.; Tarakhovsky, A. Suppression of Inflammation by a Synthetic Histone Mimic. *Nature* **2010**, *468*, 1119–1123.

203. Chung, C.-w.; Coste, H.; White, J. H.; Mirguet, O.; Wilde, J.; Gosmini, R. L.; Delves, C.; Magny, S. M.; Woodward, R.; Hughes, S. A.; Boursier, E. V.; Flynn, H.; Bouillot, A. M.; Bamborough, P.; Brusq, J.-M.G.; Gellibert, F. J.; Jones, E. J.; Riou, A. M.; Homes, P.; Martin, S. L.; Uings, J. J.; Toum, J.; Clement, C. A.; Boullay, A. B.; Grimley, R. L.; Blandel, F. M.; Prinjha, R. K.; Lee, K.; Kirilovsky, J.; Nicodeme, E. Discovery and Characterization of Small Molecule Inhibitors of the BET Family Bromodomains. *J. Med. Chem.* **2011**, *54*, 3827–3838.

204. Florvall, G.; Basu, S.; Larsson, A. Apolipoprotein A1 Is a Stronger Prognostic Marker than Are HDL and LDL Cholesterol for Cardiovascular Disease and Mortality in Elderly Men. *J. Gerontol.* **2006**, *61A*, 1262–1266.

205. Filippakopoulos, P.; Qi, J.; Picaud, S.; Shen, Y.; Smith, W. B.; Fedorov, O.; Morse, E. M.; Keates, T.; Hickman, T. T.; Felletar, I.; Philpott, M.; Munro, S.; McKeown, M. R.; Wang, Y.; Christie, A. L.; West, N.; Cameron, M. J.; Schwartz, B.; Heightman, T. D.; La Thangue, N.; French, C. A.; Wiest, O.; Kung, A. L.; Knapp, S.; Bradner, J. E. Selective Inhibition of BET Bromodomains. *Nature* **2010**, *468*, 1067–1073.

206. Borah, J. C.; Mujtaba, S.; Karakikes, I.; Zeng, L.; Muller, M.; Patel, J.; Moshkina, N.; Morohashi, K.; Zhang, W.; Gerona-Navarro, G.; Hajjar, R. J.; Zhou, M.-M. A Small Molecule Binding to the Coactivator CREB-Binding Protein Blocks Apoptosis in Cardiomyocytes. *Chem. Biol.* **2011**, *18*, 531–541.

207. Filippakopoulos, P.; Knapp, S. Targeting Bromodomains: Epigenetic Readers of Lysine Acetylation. *Nat. Rev. Drug Discov.* **2014**, *13*, 337–356.

208. Gallenkamp, D.; Gelato, K. A.; Haendler, B.; Weinmann, H. Bromodomains and their Pharmacological Inhibitors. *ChemMedChem* **2014**, *9*, 438–464.

209. Brand, M.; Measures, A. M.; Wilson, B. G.; Cortopassi, W. A.; Alexander, R.; Hoss, M.; Hewings, D. S.; Rooney, T. P. C.; Paton, R. S.; Conway, S. J. Small Molecule Inhibitors of Bromodomain-Acetyl-Lysine Interactions. *ACS Chem. Biol.* **2015**, *10*, 22–39.

210. Muller, S.; Knapp, S. Discovery of BET Bromodomain Inhibitors and their Role in Target Validation. *Med. Chem. Commun.* **2014**, *5*, 288–296.

211. Zhao, Y.; Yang, C.-Y.; Wang, S. The Making of I-BET762, a BET Bromodomain Inhibitor Now in Clinical Development. *J. Med. Chem.* **2013**, *56*, 7498–7500.

212. Shi, J.; Vakoc, C. R. The Mechanisms behind the Therapeutic Activity of BET Bromodomain Inhibition. *Mol. Cell* **2014**, *54*, 728–736.

213. Papavassiliou, K. A.; Papavassiliou, A. G. Bromodomains: Pockets with Therapeutic Potential. *Trends Mol. Med.* **2014**, *20*, 477–478.

Chapter 3

Step IIIa: Biological Hit Discovery Through High-Throughput Screening (HTS): Random Approaches and Rational Design

When one or more molecular targets are identified and validated, an R&D project enters *hit discovery (HD)*—the identification of small-molecule target modulators. HD typically entails the screening of compound collections (phases 3a and 3b, Fig. 3.1).

The pillars of HD are a reliable assay for high-throughput screening (HTS) and a chemical collection to be screened. Assay development, assay validation, hit confirmation, and prioritization pertain to the former (biology-oriented HD, phase 3a); privileged scaffolds, fragment libraries, druggable/rule of 5-compliant libraries and natural product libraries refer to the latter (chemistry-oriented HD, phase 3b, Fig. 3.1). They are described in this and the next Chapter.

A validated target has significant value, but its exploitation lies in the identification and characterization of novel, patentable small-molecule modulators. High-quality hits are the gateway to therapeutic drugs—biological candidates, albeit significant,[1] are not discussed here.

The interdisciplinary nature of HD requires multidisciplinary competence and assets.[2] Big pharma companies are suited to carry out the whole process in house, while smaller labs either rely on vendors (i.e., commercial compound collections, assay kits) and virtual approaches (i.e., computational screening and collections), or establish collaborations with labs endowed with complementary skills and equipments.

An HD effort must be well conceived and realized. As to screening campaigns, the right assay must be set up. The selected, validated target may be part of multiple biological pathways, and modulation of disease-unconnected pathways may be useless (sometimes detrimental). If so, the identified hits—and the associated costs and efforts—are wasted.

A product profile for any R&D project depends on the nature of its target(s)—membrane receptors or cytosolic enzymes, single proteins, or

Chemical Sciences in Early Drug Discovery. https://doi.org/10.1016/B978-0-08-099420-8.00003-1

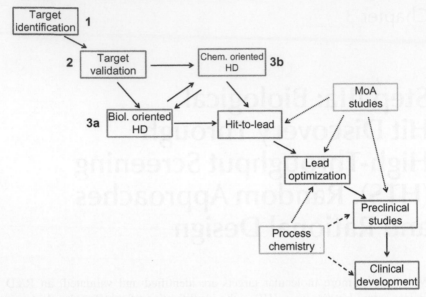

FIG. 3.1 The R&D pharmaceutical process.

multiprotein complexes. Progressable hits, selected for their activity profile, must show the potential to become drug candidates (Fig. 3.2).

A quality hit must be *chemically tractable* and *stable*, to ensure further structural optimization; it must be at least moderately *potent* against its target and *selective*, to avoid later issues about efficacy and side effects; it must be

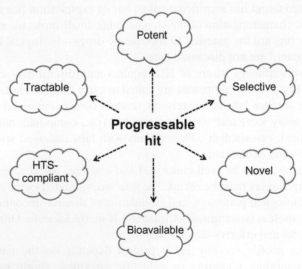

FIG. 3.2 Valuable hits: required properties.

novel per se, or transformable into novel analogues, to secure patentability; it must show acceptable *physicochemical properties* (i.e., water solubility, lipophilicity), to be bioavailable; and it must be HTS compliant—devoid of aspecific effects influencing HTS campaigns (i.e., aggregation-prone compounds, fluorescent/colored compounds).

HTS and chemical diversity (CD) collections are the foundation of R&D projects: one should attend Drug Discovery conferences to appreciate the money and efforts invested by public and private labs. Conversely, reviews on HD and early drug discovery are scarce—possibly because their complex multidisciplinary nature requires experienced researchers from big pharma to write them. Unfortunately, writing such reviews is not among their priorities.

Let us discuss how an assay for HTS is set up, and how hits are identified.

3.1. THE FOUNDATION: BIOLOGY-ORIENTED HIGH-THROUGHPUT SCREENING (HTS), NO CHEMISTRY

The discovery of penicillin[3] in late 1920s by Fleming is an example of pre-double helix drug discovery. A phenotypic effect (killing of pathogens by a micro-organism) was observed by a rigorous scientist. The following years were spent to isolate and characterize the natural product (NP) responsible for the phenotype by a multidisciplinary team at Oxford University. Penicillin was a therapeutic breakthrough, but its discovery process was slow and inefficient.

Once the genetic code was unraveled in the 1960s,[4] a switch between phenotypic observations and target-based, rational drug design was started. The following decades saw the gradual disappearance of low-throughput, in vivo screening and the emergence of in vitro screening. -Omic sciences were applied to target identification[5] and validation[6]; structural sciences provided information on target structure and mode of action,[7] enabling in silico modeling[8]; cellular and molecular biology ensured the expression and purification of putative targets,[2] leading to cell-free and cellular HTS assays.[9]

In vitro screening provides multiple advantages vs. in vivo testing. They are illustrated in Fig. 3.3.

In vitro assays reduced the biological burden (no animals/*ethical issues*). Assays were run in parallel in the 1980s, using early versions of microtiter plates and instrument readers.[10] *Miniaturization* and *standardization* resulted from the introduction of 96-well microplates (well volume: 100–200 µL) in early '90s[11]; the definition of microplate standards for commercial microplates and instruments in late 1990s[12]; the definition of HTS standards[13]; the gradual replacement of 96-well microplates with 384- and 1536-well plates (well volume of 30–100 and 2.5–10 µL, respectively)[14]; and the release of best-practice manuals for HTS assay development and execution.[15]

Modular *automation*[16] increased the throughput of HTS from a few hundred compounds/day/person in early 1990s to ≈50 K compounds/day/person in late 1990s,[17] to ultra-HTS/>100 K compounds/day/person now.[18] The *cost*

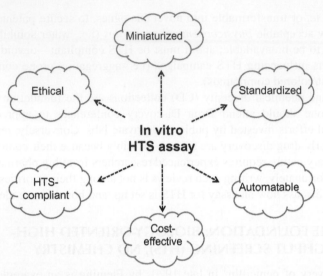

FIG. 3.3   HTS assays: required properties.

of modular automation gradually declined, and is now affordable for academic labs and small biotechs.[19] *Robustness* and *compliance* to HTS characterize reliable, target-specific in vitro HTS assays (Fig. 3.3).

Smaller 3456-well plates (well volume: 1–2 µL) were introduced,[20] but working at the boundary between micro- and nanoplates changed key parameters (i.e., volume to surface ratios) and affected assay reliability.[12] Microfluidics[21,22] increased miniaturization and were used for analytical applications. Their usage in the HTS field was limited, due to technical issues (i.e., nonphysiological behavior of cells, compound/reagent delivery). Miniaturization to the nL level is not necessary, as the reduction in compound/reagent/costs with respect to existing 10–50 µL-based HTS platforms would be limited.[12]

In early 1990s, HTS mostly relied upon *cell-free/target-based assays*[23] (Fig. 3.4, *bottom left*), based on the postgenomic wave of identified targets.[24] Screening protocols were compatible with target handling in 96-well plates, and with medium-large screening sets (up to ≈200 K compounds) screened in a few weeks.[17] The following decade saw an increase of *cellular/phenotypic assays*[25] (Fig. 3.4, bottom right), due to simpler HT cell manipulation, and to suboptimal performances of target-based assays.[26] Cellular assays were heavily represented in HTS since mid 2000s,[27] and *whole organism assays* (zebrafish, worms, insects, plants) (Fig. 3.4, *top*) became popular to select hits for their metabolism and safety profile.[28]

Target-, cell-, and whole organism-based assays ideally should be deployed in parallel to find good hits—if the primary HTS assay is target based, the profiling/secondary assay should be cell based, and vice versa.[29] Target-based discovery entails a limited cost of biologicals and chemicals, and its throughput is

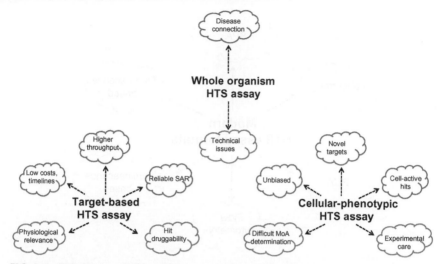

FIG. 3.4   HTS assays: cell-free, cellular, and whole organism assays.

high. It measures the affinity for a chosen target, leading to accurate structure-activity relationships (SARs) for compound classes, and to the rational design of optimized analogues.[30] Conversely, hits from target-based HTS may be inactive in a cellular model due to their properties (i.e., poor physicochemical properties, instability), or to the choice of a disease irrelevant target.[31]

Cellular/phenotypic assays are unbiased and should lead to the identification of unknown targets in complex diseases (Chapter 1). Hits from phenotypic assays are bioavailable and stable in cellular environments.[31] Conversely, the molecular target-hit connection is tough to unravel (see Chapter 2), the experimental conditions must be carefully controlled, and the rational design of optimized hits is more challenging.[5] A higher throughput and lower costs/timelines are achievable with target-based HTS, but cellular HTS is closing the gap.

Whole organism assays are the long-term evolution of HTS, having superior significance in terms of disease recapitulation and bioavailability. Issues in terms of formats, throughput, timelines, and costs are limiting their wider usage now, but the gap with other HTS options will be closed.

Homogeneous mix and split HTS assay formats and detection techniques, suitable for target-based and cellular assays, enabled HTS on most validated targets.[18] Separation-based enzyme-linked immunosorbent ELISA assays were replaced by formats shown in Fig. 3.5.

Radioactivity detection is extremely sensitive, but almost disappeared from HTS due to waste disposal and safety problems. Scintillation proximity assays/SPA (target-based) using "safe" [3]H isotopes are a notable exception.[32] *Fluorescence detection* is available in various assay versions. *Direct fluorescence* measurements detect reaction quenching (target- and cell-based),[33] or measure kinetic/concentration (i.e., fluorescent-imaging plate reader/FLIPR[34])

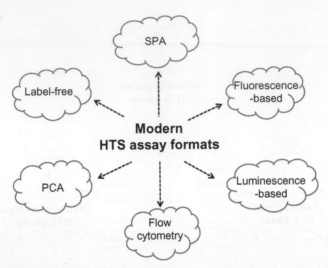

FIG. 3.5　HTS assay formats: radioactivity-, fluorescence- and luminescence-based, flow cytometry, PCA, and label-free formats.

or localization changes (i.e., high content screening/HCS[35]) in cell-based formats. *Indirect fluorescence* measurements detect protein-ligand complexes when stimulated with polarized light (target-based fluorescence polarization/FP[36]), or the energy transfer between two fluorescence molecules (target- and cell-based fluorescence resonance energy transfer/FRET,[37] homogeneous time-based time resolved fluorescence/HTRF[38]). *Luminescence* is a target- and cell-based format used in bioluminescence resonance energy transfer/BRET[37] (energy transfer between luminescent molecules), in amplified luminescence proximity/AlphaScreen[39] (detection of a chemiluminescent reaction caused by singlet oxygen), and in electrochemiluminescence[40] (detection of an electroluminescent reaction/excited electronic state). Popular target- and cell-based methods include *protein complementation assays/PCA*[41] (interaction between two protein halves) and *flow cytometry*[42] (particle isolation based upon size). *Label-free technologies*[10,43] use biophysical detection (i.e., calorimetry, refraction index, calorimetry) in target- and cell-based assays to avoid any interference by labeled reagents, although other inconveniences were reported.[18]

HTS has become an established approach to identify hits in pharmaceutical companies around the year 2000. Technological advancements and cost effectiveness have made HTS available in academic centers.[44] The upgrading of academic HTS was due also to experts from big pharma HTS groups being laid off and joining academic groups.[18]

Private and public HTS facilities often differ in terms of targets and expectations. Companies aim to discover chemically tractable hits to be converted through extensive (but affordable) structural optimization into leads, candidates, and drugs. Public labs are attracted by the scientific aspects related to HTS, and often look for small-molecule chemical probes[45]—we talk more about them and chemistry-related topics in HD in the next Chapter. As to targets, private HTS campaigns are directed against validated targets to minimize the risk of nondevelopable hits. Public efforts are often based upon phenotypic screenings, to elucidate novel mechanisms of putative therapeutic intervention.

How does HTS fare in terms of success, as a tool to improve drug discovery (companies—drug-like hits) or to further clarify pathological mechanisms (academia—chemical probes)? Recent reviews[46,47] blame HTS for the slightly decreased number of yearly approved drugs: an increased number of tested compounds, they say, should have positively impacted the end results of pharma R&D projects. High costs, lack of hits from ≈50% of HTS campaigns, loss of creativity among researchers, and poor quality of results were used to define HTS an expensive, almost useless toy for technologists. Experts from industrial and academic HTS groups recently[48] rejected these (according to them) unsubstantiated critiques by claiming high quality in HTS design and experimentation (poor quality is chosen by less-than-rigorous scientists, as in any research area); by defining a limited 10%–25% time and cost increase for pharma R&D projects using HTS on large compound collections (long timelines and high costs are typical for drug discovery, and the blame should not be put on HTS); by requiring intellectual neutrality for HTS (higher throughputs should not decrease the intellectual input of researchers, which have more data to select hits/progress projects); and by asserting a ≥ 60% applicability for HTS, better than any other hit-finding approach (unsuccessful attempts are due to the untractable nature of a few targets, rather than to HTS).

Is HTS in drug discovery a key asset toward novel drugs, a fancy and expensive irrelevant toy, or anything in between the two? The next Section illustrates two modern methods that support the discovery of quality hits in a cost-, effort-, and result-effective manner.

3.2. VIRTUAL HIGH-THROUGHPUT SCREENING (VHTS)

Costs related to the discovery of drug candidates, and to their development up to marketed drugs are approaching the 3-billion dollar mark.[49] Costs associated with tangible HD include the acquisition/synthesis of diverse chemical collections; the development and automation of disease-related HTS assays; the acquisition and maintenance of an HTS platform, including compound handling (logistics, quality control, etc.), HTS instrumentation (sample preparation, assay modules, detection/readout, etc.), and IT management (data acquisition and storage, database organization, etc.).

Computer-based methods gained importance in pharmaceutical R&D in the last decades[50]—for example, HT genome browsing to detect mutations and polymorphisms (bioinformatics[51]), analysis/definition of pathophysiological networks (systems biology[52]) and of compound bioavailability (pharmacokinetic/pharmacodynamic modeling[53]).

Computational tools do not allow us to reproduce in silico/on a computer screen a cellular assay: the number of interactions among molecular entities taking place in a single cell at a given time exceeds the computing power of supercomputers. Conversely, *computer-assisted drug design (CADD*[54]*)* predicts in silico the putative interactions between a structurally characterized target and a small-molecule effector in an artificial, isolated environment. The following pages are an end-user-oriented description of computational tools assisting HD. Listed references provide readers with details about CADD principles and applications in drug discovery.

Target structures cannot be generated in silico without experimental data, due to their complexity. Fig. 3.6 summarizes the creation of a virtual target structure from abundant (top left) or limited (top right) experimental data.

The use of X-ray crystallography,[55] nuclear magnetic resonance (NMR)[56] and/or cryoelectron microscopy (cryo-EM[57]) to determine the structure of a pure protein has been reviewed.[58] Raw data from one or more sources must be refined to create a reliable in silico target model.[59] Tautomeries must be resolved; ionization states must be determined; binding site(s) must be determined; hydrogen atoms, water molecules, metal ions, activators, and/or cofactors must be added, in a target- and software-specific manner.[60]

Computational tools help if scarce experimental data about the target cannot *per se* define its structure, but can guide de novo *modeling*[61] (e.g., NMR nuclear Overhauser effects/NOE,[62] electron paramagnetic resonance/EPR[63]); if the target is homologous (at least ≈40% primary sequence, higher

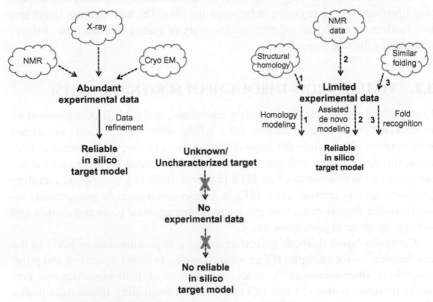

FIG. 3.6   Virtual models of molecular targets: different scenarios.

homology in the binding site[64]) to a structurally characterized protein—*homology modeling*[65]; and if one or more structurally characterized, non-homologous proteins show similar folding patterns—*fold recognition.*[66] Technology advancements notwithstanding, an in silico model for some therapeutically relevant targets—e.g., most membrane proteins[67]—is not yet achievable (*bottom*, Fig. 3.6).

As to small-molecule ligands, their virtual models require in silico representation and determination of their properties. The former depicts the preferred, energy-minimized conformations of a ligand. The latter—i.e., the blockbuster atorvastatin—stems from its 1D (e.g., molecular formula and mass, *top*, Fig. 3.7), 2D (e.g., rotatable bond number, middle), and 3D features (e.g., polar surface area/PSA, bottom, Fig. 3.7).

Properties are converted into numerical values named *molecular descriptors.*[68] Molecular descriptors can be measured (experimental descriptors: e.g., dipole moments, partition coefficients, melting points) or calculated (theoretical descriptors: e.g., structural fragments, fingerprints).[69] >3000 molecular descriptors are known,[70] and many softwares,[71] based on their combinations

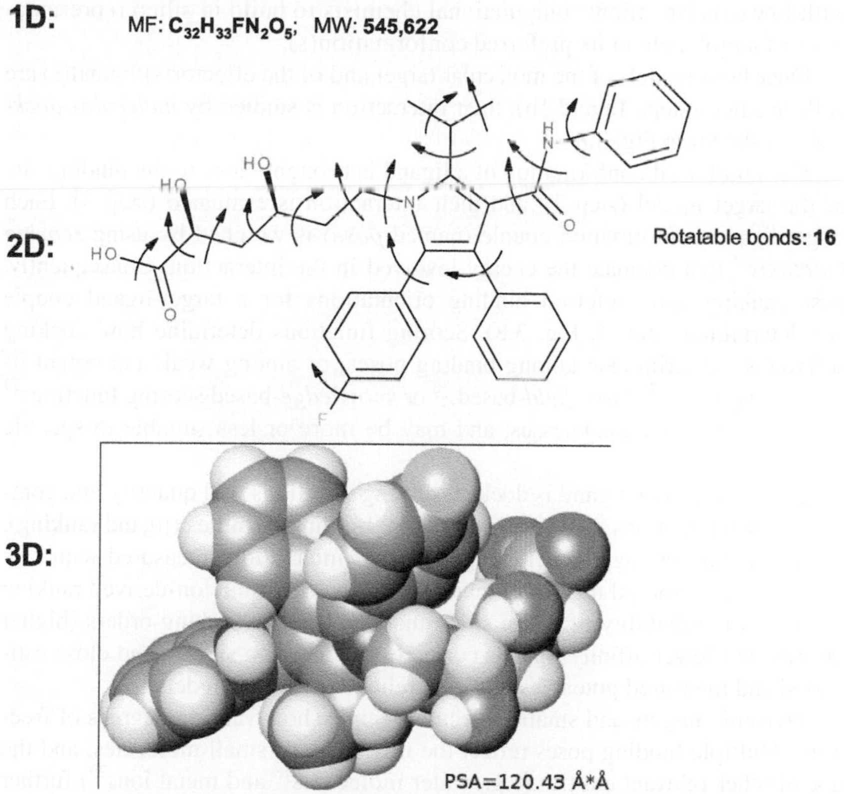

1D: MF: $C_{32}H_{33}FN_2O_5$;   MW: **545,622**

2D: Rotatable bonds: **16**

3D: PSA=120.43 Å*Å

FIG. 3.7   1D-, 2D-, and 3D-features of virtual small molecules.

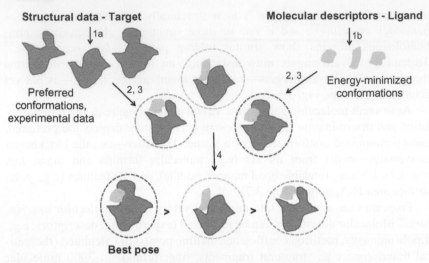

Structural data - Target

Molecular descriptors - Ligand

Preferred conformations, experimental data

Energy-minimized conformations

Best pose

FIG. 3.8   Molecular docking of putative binders: pose creation and ranking.

with low overlap, allow computational chemists to build in silico representations of a molecule in its preferred conformation(s).

Once both models of the molecular target and of the effector(s)/ligand(s) are built in silico (steps 1a and 1b), their interaction is studied by *molecular docking*[72] as shown in Fig. 3.8.

Each preferred conformation of a ligand is brought close to the binding site of the target model (step 2), and their interaction is evaluated (step 3). Each target-ligand conformation couple (named *pose*) is weighed by using *scoring functions*[73] that estimate the energy involved in the interaction. Consequently, pose ranking and preferred binding orientations for a target-ligand couple are determined (step 4, Fig. 3.8). Scoring functions determine how docking softwares[74] discriminate among binding poses, or among weak and potent ligands. *Empirical*,[75] *force-field*-based,[76] or *knowledge*-based scoring functions[77] have strengths and weaknesses, and may be more or less suitable to specific applications.

If more than one ligand is docked, scoring functions will quantify and compare their interactions, as a surrogate of real binding affinities (ligand ranking). Experimental binding potencies (ligand-target interactions measured with biological assays) should then be compared with scoring function-derived ranking to check the reliability of the in silico models. Similar ranking orders (higher affinity and lower affinity ligands correctly predicted in silico), and close estimated and measured potencies, indicate reliable in silico models.

Dynamic targets and small-molecule ligands show varying degrees of freedom. Multiple binding poses reflect the flexibility of small molecules, and the use of other relevant entities (e.g., water molecules[78] and metal ions[79]) further improves the binding mode reliability. An in silico evaluation should consent target flexibility to adopt multiple conformations (i.e., multifunctional proteins, one function/one conformation[80]); and/or to adapt to their binding partner's preferences (i.e., slightly rearranging the binding site geometry to strengthen the interaction with a ligand[81]). Unfortunately flexible docking approaches are computation time intense, and a compromise between rigorousness and manageable timelines must be found.

In silico protein flexibility was obtained by allowing small target-ligand overlaps before energy minimization (*soft docking*[82]), or by including aminoacid *side chain flexibility*/rotamer library sampling.[83] Both methods are not too time consuming in terms of calculations, but their sampling of protein flexibility is limited. *Ensemble docking*[84] uses multiple protein conformations, either from biophysical measurements (e.g., crystals of cofactor- or effector-bound complexes[85]) or from dynamic simulations on a single available conformation (e.g., enhanced sampling via temperature-accelerated dynamics[86] or metadynamics[87]). Ensemble docking is better than rigid docking,[88] but computational needs and longer simulations prevent its widespread usage.[89]

Let us now move to HD. Once the HD phase kicks off, four scenarios depicted in Fig. 3.9 may happen.

If the molecular target is structurally not characterized, and its ligands are unknown (*bottom left*, Fig. 3.9), CADD cannot be deployed—target sequence-derived models[90] are limited in scope.

If an in silico model of the target is available and endogenous ligands are unknown (*bottom right*, Fig. 3.9), *structure-based drug design (SBDD*[59]) is used to look for small-molecule ligands.

The drug-like chemical space for small molecules was estimated between 10^{24} and 10^{30} virtual compounds.[91] Such numbers largely exceed the throughput of synthetic efforts. How to choose the right ligands for a given target, then? Working in silico cancels synthetic issues, but does not reduce a priori the size of the chemical space.

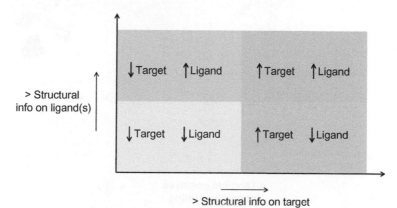

FIG. 3.9   Information-driven virtual screening: different scenarios.

The in silico model of a small molecule is built via molecular descriptors and softwares, as described earlier.[70,71] Large *virtual libraries* made by n drug-like small molecules are built by repeating the process n times. Preassembled virtual libraries containing millions of compounds can be downloaded free of charge from public sites,[92,93] and may be enriched with proprietary compounds. CD-based libraries sample the small-molecule chemical space, while focused virtual libraries[94] (e.g., kinase-directed[95]) sample parts of the CD space. Libraries may be filtered to discard nondrug-like compounds (reactive, toxic, assay interfering, aggregation prone),[96–98] or may be compound source specific (i.e., natural product,[99] or macrocyclic libraries[100]). Library filtering/assembling may be done using available online tools for library design.[59] Large virtual libraries containing millions of drug-like molecules, and/or smaller time- and cost-effective virtual collections[101] can be assembled and used to assist the HD process.

Structure-based *virtual high-throughput screening* (*vHTS*[8]) selects hits from a virtual library tested against the model of a target (Fig. 3.10).

After in silico loading of the target model (step 1, Fig. 3.10), and of the preferred conformations for each virtual library individual (step 2), the best binding pose is selected for each target-fitting library individual (step 3) while the majority of nonfitting individuals are discarded (step 4). Fitting poses are then scored for their putative target affinity (step 5), and target-binding library individuals are ranked accordingly (step 6). The highest-ranked library individuals are then either purchased (commercially available) or synthesized to validate

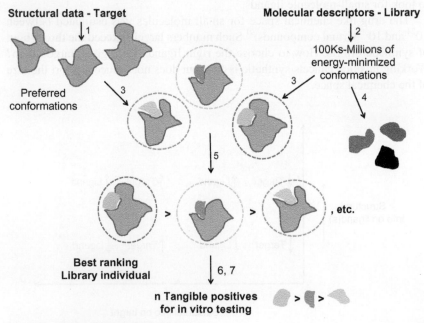

FIG. 3.10   Structure-based vHTS: the process.

the results of vHTS (step 7, Fig. 3.10; typically $20 < n \langle 200 \rangle$). Tangible positives are then tested in a target-based biological assay to yield validated hits and to determine the performance of the docking-based vHTS campaign. An enrichment with respect to the <1% random HTS hit rate is expected; if that does not happen, the model of the target needs major refinements.

Reliable molecular descriptors and softwares could precisely estimate the putative biological activity of all library individuals in vHTS, but wastage of time and money must be avoided. Rigid docking (single-target conformation, limited ligand flexibility) is used in a first vHTS step to rapidly discard the vast majority of library individuals; calculation-demanding docking protocols are then used on few positives (multiple target and ligand conformations, water, metal ions, etc.). False positives are prevented, and a reliable ranking is obtained with limited efforts.[8] A structure-based *pharmacophore*[102] is sometimes generated by clustering putative interaction zones in the target-binding site in a tridimensional set of electronic and structural features needed to interact with the target.

SBDD and docking cannot be deployed if a virtual target model is unavailable, but CADD is useful in HD if several potent ligands are known (Fig. 3.9, *top left*). *Ligand-based drug design (LBDD)*[103] estimates the similarity between virtual library individuals and extrapolated models of known ligands. An indirect estimation of target-binding affinity can be obtained via *similarity searching,*[104] i.e., comparing known ligands with a virtual library of molecules. Similar compounds should show similar biological activities, as ligand-based similarity searches led to the identification of active hits.[105]

Similarity searches are less computing and time consuming than docking, as they do not entail complex protein structures. Nonetheless, ligands can be converted into models with varying complexity; the most accurate descriptions correspond to longer calculations and higher prediction accuracy.[104]

2D-Molecular fingerprints[106] can be calculated for known ligands, and compared with the fingerprints of each individual in a virtual library. They are made of strings of binary bits, representing some molecular features that may be present (1-yes) or absent (0-no) in a compound. Their performance is heavily dependent on the relevance of such molecular features in a virtual library described using them.[105] Fingerprints made by longer strings (up to >2000 features[107]) better describe compounds and compare them.

Substructure-based fingerprints[108] are the simplest, most used 2D fingerprints. They define a set of chemical substructures and look for their presence in small molecules. Their strings vary between ≈100 and > 1000 bits (sometimes they are customizable[109]). They are target dependent—if their set of substructures is poorly represented in known ligands, another fingerprint should be used. A 16-bit substructure fingerprint portion for atorvastatin (*top*) and rosuvastatin (*bottom*) is shown in Fig. 3.11.

Topology-based fingerprints[110] are assembled by calculating the unique paths (mostly linear, up to a predefined length) from each atom in a molecule.

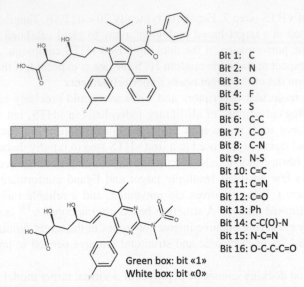

Bit 1: C
Bit 2: N
Bit 3: O
Bit 4: F
Bit 5: S
Bit 6: C-C
Bit 7: C-O
Bit 8: C-N
Bit 9: N-S
Bit 10: C=C
Bit 11: C=N
Bit 12: C=O
Bit 13: Ph
Bit 14: C-C(O)-N
Bit 15: N-C=N
Bit 16: O-C-C-C=O

Green box: bit «1»
White box: bit «0»

FIG. 3.11 Substructure-based 2D-fingerprints: atorvastatin and rosuvastatin.

They do not depend on specific substructures, and may cause bit collisions (multiple paths leading to the same bit), especially when their string/bit number is short. An example of a 10-bit topological fingerprint covering 1- to 7-bond linear paths from the terminal OH group of atorvastatin is reported in Fig. 3.12.

In *circular paths*,[111] an atom is the center of a connectivity table with 0 to n atom radii levels. Such fingerprints cannot be used for substructure searching, but perform well in full structure similarity searching. The circular connectivity/fingerprint of atorvastatin (starting from the benzylic carbon, up to 3-bond circular paths) is shown in Fig. 3.13.

Hybrid fingerprints[112] partition the calculated value for a compound descriptor (typically ≈100 descriptors) in a predetermined range (typically 2–7 intervals), and fill the binary bit in accordance with such value. The number of filled/1/present bits for each library individual is constant, and the descriptor ranges are customizable—see Fig. 3.14 (PSA, PubChem values, three statins).

3D-Molecular fingerprints[113] describe molecular features of virtual library individuals from their 3-D representations. They rely on "an ensemble of steric and electronic features that is necessary to ensure the optimal supramolecular interactions with a specific biological target and to trigger (or block) its biological response",[114] i.e., a 3D-*pharmacophore*. A 3D-pharmacophore is extracted from the minimized conformations of ligands. It is made of 3–4 common structural features (i.e., lipophilic, hydrogen bond-HB donor or acceptor, positively or negatively charged groups) that should determine their activity on the target, and the characterization of their spatial arrangements with some tolerance (in a 1–2Å range, considering conformational flexibility of the ligands[115]). A 3-point

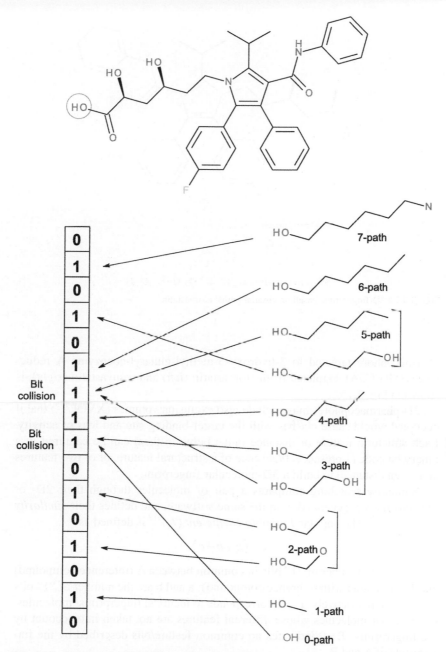

FIG. 3.12   Topology-based 2D-fingerprints: atorvastatin.

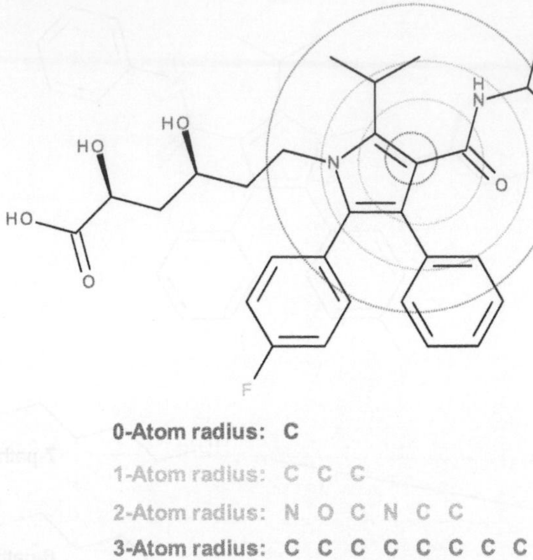

0-Atom radius: C

1-Atom radius: C C C

2-Atom radius: N O C N C C

3-Atom radius: C C C C C C C C

FIG. 3.13   2D-fingerprints based on circular paths: atorvastatin.

pharmacophore targeted to 3-hydroxy-3-methyl-glutaryl-coenzyme A reductase (HMG-CoA), extracted from atorvastatin (*left*) and rosuvastatin (*right*), is shown in Fig. 3.15.

3D-pharmacophores may include also exclusion volumes (XVOL[116]) that if occupied would cause clashes with the target-binding site and loss of activity. Each structural feature or distance range between them can be captured in a binary bit code (1 presence, 0/absence of a structural feature, or of two features at a given distance) to build a 3D-molecular fingerprint.

Similarity searching compares a pair of molecules through their 2D- or 3D-fingerprints, generated with the same software, and defines their *similarity coefficient.*[117] The popular *Tanimoto coefficient* (T_c)[118] is defined as

$$c \div (a + b - c),$$

where c is the number of "1" bits in common between A (reference compound) and B (compared with reference compound); a and b are the number of "1" bits in A and B, respectively. $T_c = 1$ corresponds to identical fingerprints/ molecules, or to similar molecules whose different features are not taken into account by the fingerprints; $T_c = 0$ indicates no common feature/bits described by the fingerprints of A and B.

Fingerprint-based similarity searches are more accurate using longer bit strings/compared features, or including "0" bits/absence of a molecular feature in the comparison.[119] Both approaches increase the reliability of T_c values, but require longer computing time.

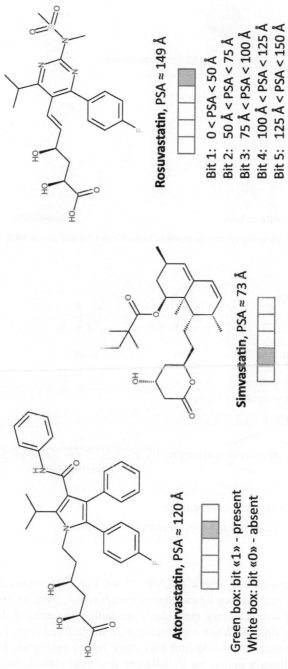

Rosuvastatin, PSA ≈ 149 Å

Bit 1: 0 < PSA < 50 Å
Bit 2: 50 Å < PSA < 75 Å
Bit 3: 75 Å < PSA < 100 Å
Bit 4: 100 Å < PSA < 125 Å
Bit 5: 125 Å < PSA < 150 Å

Simvastatin, PSA ≈ 73 Å

Atorvastatin, PSA ≈ 120 Å

Green box: bit «1» - present
White box: bit «0» - absent

FIG. 3.14   Hybrid 2D-fingerprints: atorvastatin, simvastatin, and rosuvastatin.

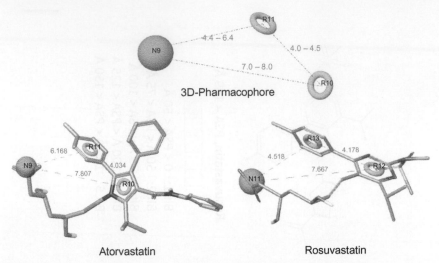

FIG. 3.15 3-point pharmacophores: an example from atorvastatin and rosuvastatin.

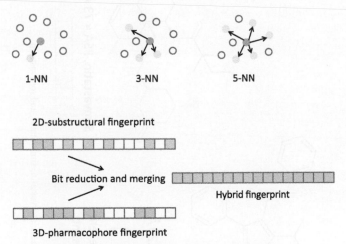

FIG. 3.16 Nearest-neighbor searches and fingerprint recombination.

The *nearest-neighbor* search (n-(NN),[120] see Fig. 3.16, *top*) calculates T_c coefficients for each known ligand-library member pair; then, an average T_c is calculated, and virtual library individuals are ranked in terms of their overall similarity to the ligands. When n is small (i.e., a single filled circle / ligand, *top left*), similarity searches should find only close ligand analogues. For larger n values (i.e., 5 ligands belonging to different structural classes, top right), the search should lead to novel, patentable chemotypes (*scaffold hopping*[121]).

Fingerprint recombination (Ref. 122 Fig. 3.16, *bottom*) entails the reduction of bits in significantly fingerprints (2D- and 3D-fingerprints in Fig. 3.16), choosing only reference compound-relevant ones, and merging the remaining bits in a hybrid fingerprint.

Similarity searches were reviewed.[104–106,123] Drawbacks include dependence on compound classes (any fingerprint type performs well on some chemotypes, and poorly on others[105]); the lack of reliable thresholds for similarity coefficients to select similar compounds (the proposed $T_c=0.85$ value sometimes has a $<30\%$ probability to identify an active compound[124]); the poor correlation between fingerprint complexity and quality of results (2D-based, less complex/time consuming fingerprints may perform better than more complex 3D fingerprints[125]); a moderate predictivity for fingerprint-based similarity searches ($<20\%$/a few actives in 100 best ranking virtual hits are to be expected[126]). Despite all, similarity searches by molecular fingerprints provided good results in virtual HD. They are easily used, as open source platforms are publicly available on the Web,[127] and computing time/efforts are limited. *Dissimilarity searches*[128] are important in HD, when virtual or tangible compound collections must be assembled for vHTS or HTS, respectively. Dissimilarity-based compound selections will be described in detail in the next Chapter.

A computer-intensive, ligand-based strategy relies on previously mentioned *pharmacophores* (Figs. 3.17 and 3.18).

A set of known target ligands and of their preferred conformations (*training set*) is built in silico (step 1a, Fig. 3.17). The training set is used to extract a pharmacophore model (step 2). Rather than reducing it to bit strings, the

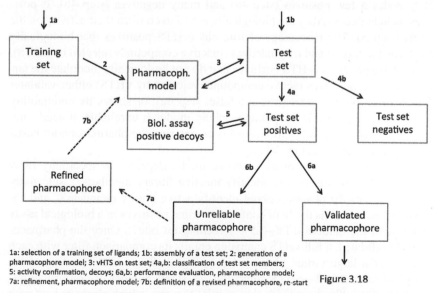

1a: selection of a training set of ligands; 1b: assembly of a test set; 2: generation of a pharmacophore model; 3: vHTS on test set; 4a,b: classification of test set members; 5: activity confirmation, decoys; 6a,b: performance evaluation, pharmacophore model; 7a: refinement, pharmacophore model; 7b: definition of a revised pharmacophore, re-start

Figure 3.18

FIG. 3.17   3D-pharmacophore searches: from training sets to validated pharmacophores.

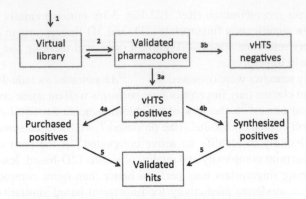

1: selection of a large virtual library; 2: vHTS campaign; 3a: selection of fitting positives; 3b: elimination of inactives; 4a: purchase of commercially available positives; 4b: synthesis of unavailable positives; 5:

FIG. 3.18 3D-pharmacophore searches: from validated pharmacophores to validated hits.

3D-pharmacophore is validated and used as a vHTS model as such.[129] Validation requires the in silico creation of a *test set* (step 1b). The test set contains known ligands not included in the training set; known inactives (negative controls); and decoys (unknown activity on the target) with similar physicochemical properties to known ligands. Decoys can be assembled from a Web-based directory of useful decoys (DUD[130]). Inactives and decoys should exceed known actives (at least ≈50:1), to simulate a vHTS campaign, as virtual libraries contain many more inactives than actives. A vHTS campaign on the test set (step 3) provides a few positives (step 4a) and many negatives (step 4b). If positives include decoys, they are biologically tested to confirm their activity on the target (step 5). The percentage of true hits (vHTS positives that biologically modulate the target) and true inactives (inactive compounds rejected by vHTS), versus false positives (vHTS positives that do not biologically modulate the target) and false negatives (active compounds rejected by vHTS) either validates the pharmacophore model (true >> false, step 6a), or shows its unreliability (true ≤ false, step 6b). If the pharmacophore model is unreliable, it needs major refinements (step 7a) before possibly re-entering the pharmacophore-based vHTS process (step 7b, Fig. 3.17).

The predictivity of a pharmacophore model depends on specificity (how well does the pharmacophore identify inactive library members?), selectivity (how well does the pharmacophore identify active library members?), and enrichment factor/EF (is the % of confirmed virtual positives in a biological assay higher than the random 0.1%–0.5% screening hit rate?). Once the pharmacophore is validated, a full vHTS campaign entails its comparison filter with each member of a large virtual library (step 1, Fig. 3.18—in silico loading of the virtual library; step 2—vHTS campaign). A small number of conformations/flexibility for a library individual may miss the one, which aligns at best with the pharmacophore—*false negative*; too much flexibility may create fitting conformations, which do not exist in biological environments—*false positive*.[131] Million-membered libraries do not allow high compound flexibility, due to computing time limitations. The alignment between a pharmacophore model and library members can be evaluated via root mean square deviation (RMSD)-based methods (measuring the distance between the centre of pharmacophoric groups and their counterpart in the aligned library individual)[132]; overlay-based methods (centers are substituted by radii)[133]; and shape-based methods (shape-based comparisons between pharmacophores and library individuals).[134] A detailed description among eight algorithms/softwares showed their strengths and weaknesses, and the selectivity, specificity, and EF obtained on four molecular targets.[135]

Pharmacophore-based comparisons are expressed as *fit values* between the model and each library member reasonably aligning with the pharmacophore. Fitting positives are selected (step 3a), while inactives are discarded (step 3b). Fit values are calculated comparing each pharmacophoric feature and its counterpart in fitting positives in terms of chemical class and relative orientation. Weighing coefficients may be attributed to each feature, to account for their importance in the interaction with the molecular target.[129] Higher fit values indicate a better alignment than lower ones, but other factors (i.e., degree of ligand flexibility in the calculation, number and drug-likeness of structurally similar positives) influence their prioritization for purchase/synthesis (respectively step 4a and 4b). Tangible positives are then biologically tested to yield a few validated hits (step 5, Fig. 3.18), and to evaluate the performance of the pharmacophore-based vHTS campaign.

An ideal vHTS scenario entails a structurally characterized molecular target and several validated ligands (Fig. 3.9, *top right*). Rigorous structure-based protocols are time consuming and require computing power; conversely, ligand-based methods are less demanding but often not so predictive. A good compromise involves an initial similarity search to rule out the vast majority of inactives, followed by structure-based characterization/validation of positives using rigorous docking protocols.[136]

If available structural data include one or more X-ray target-ligand complexes, *structure-based pharmacophores*[102] should be considered. Their generation is simple, as the arrangement of a ligand in the binding site can be rapidly extrapolated into a model. Multiple complexes between the target and several ligands should, after superimposition and removal of overlapping features, lead to more predictive merged pharmacophores.[137] Comprehensive, structure-based pharmacophores recapitulated information from >100 different target-ligand complexes.[138] The results of a vHTS campaign based on them—still significantly less time-consuming than docking—could be refined with rigorous docking on the small number of virtual positives.[102]

The choice among SBDD- and LBDD-based methods to screen a virtual library depends on the available structural information (Fig. 3.9), on the accessible hardware and software, and on the R&D project timelines. Each among them leads to virtual positives that must be prioritized, devirtualized, and validated as tractable hits.[139–141] The success of a vHTS campaign depends on positives, that may be refined pre- (i.e., chemical and structural filters) or post-vHTS (i.e., clustering of positives). We'll discuss more about positives, hits, chemical filters, and compound collections (virtual for vHTS, tangible for HTS) in the next chapter.

3.3. BIOPHYSICAL SCREENING IN FRAGMENT-BASED DRUG DISCOVERY

HTS was criticized for its shortcomings,[46,47] due to unreasonable expectations and scarce reports about HTS campaigns in successful R&D projects. Heated topics for debate involve HTS cost (large numbers of compounds and assays, extensive automation) and reliability (false positives and negatives in HTS campaigns, large variations). Two apparently unrelated solutions—biophysical screenings and smaller molecules/fragments—merged into an effective approach toward the discovery of progressable hits from HTS.

Drug-like/bioavailable small molecules were characterized by Chris Lipinski's *rule of five*[142]—see the next Chapter for more details. Any such small molecule may be the right *Schlussel* (key) in a molecular target's *Schloss* (lock), as defined by Emil Fischer in 1894.[143] We discussed the 10^{24}–10^{30} Lipinski rule-respecting, synthetically accessible small molecules.[91] Even if the biologically relevant chemical space may be smaller,[144] HTS collections made by $\geq 1\,M$ compounds represent a tiny portion of such space. How then to increase the representation of the biologically relevant chemical space by an HTS collection (higher probability of success in finding progressable hits) while reducing its size (lower costs and efforts)?

A perfect fit requires multiple interactions among the teeth, the notches, and the grooves on a key blade, and the inner structure of a lock. A single mismatch prevents the fit and the opening of the door. The same is true for a ligand-target interaction. Molecules with ≈30 heavy atoms contain multiple interactors (HB donor and acceptors, ionized groups, aromatic rings, etc.) in a fixed spatial arrangement. Target-binding sites may be large or small, shallow or deep, but always defined by precisely oriented aminoacidic residues. They accommodate the ligand structure, and any repulsive interaction (steric clash, electrostatic repulsion, etc.) must be avoided. The fit between a 5-feature ligand and its target-binding site is shown in Fig. 3.19, *top left*. A single mismatch (*bottom left*, modified shape) overcomes the four matching interactions and prevents binding.

A few randomly selected drug-like molecules in an HTS collection fits with the binding site, and shows affinities as hits (typically 0.1%–0.5% of the screening collection, with low μM potency in biological assays).

Let us introduce the master key. Its teeth-notches-grooves profile is simplified, and works on several locks; the simplification produces no mismatches/clashes, preserves 1–2 key interactions and leads to door opening. If we switch

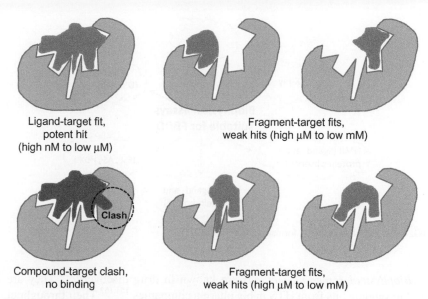

FIG. 3.19   Ligand-target and fragment-target fits: matches and mismatches.

to ligand-target complexes, a smaller molecule (*fragment*, MW ≤ 300 Da, 15–17 heavy atoms on average, compared to 28–30 heavy atoms in small molecules) establishes 1–2 interactions with a target and easily fits with a binding site. Fig. 3.19, top and bottom, shows four fragments that fit a binding site in a graphic illustration of *fragment-based drug discovery (FBDD)*.[145–147]

A comprehensive database of drug-like fragments up to 17 heavy atoms contains ≈166.4 billion compounds (GDP-17[148]). An order of magnitude in size increase for each added heavy atom leads to ≈10^{28} individuals for 30 heavy atom-containing libraries, in accordance with.[91] Thus, smaller fragment collections (a few hundreds to a few thousands, see next Chapter) better recapitulate the biologically relevant CD space, while reducing costs for library generation, logistics, and screening.

Fragment screening requires only MT (medium-throughput) screening, but its sensitivity must increase to detect the weaker (typically high μM-low mM) fragment-binding site interactions. Assay formats suitable for FBDD are listed in Fig. 3.20.

Biochemical HTS assay formats working close to mM concentrations (*high concentration screening-HCS*[149]) increase the occurrence of false positives (i.e., compound aggregation, disturbance of biochemical environments, interference with readout methods) and false negatives (i.e., compound precipitation). HCS does not provide structural information about the target-ligand interaction. Nevertheless, technology improvements and HT capacity make HCS a viable choice for FBDD.[146]

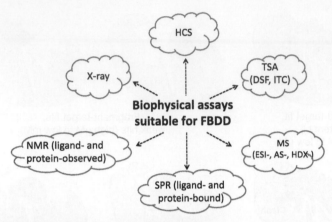

FIG. 3.20   Biophysical assay formats suitable for FBDD.

Biophysical screening methods are known in drug discovery.[150] They are used to validate hits from HTS in big pharma companies.[151,152] Their throughput is lower than biochemical assays, although HTS campaigns using mass spectrometry (MS)-based,[153] thermal shift assay (TSA)-based[154] and NMR-based[155] assays were reported. Their deployment on HTS positives confirms their binding with the target, validates them as true hits, and provides structural information regarding the ligand-target interaction.

Fragment collections vary between 500 and 20,000 individuals (typically in the 1000–5000 range).[147] Several biophysical techniques have a suitable throughput as primary assays, while others can be used as secondary/confirmation- and information-gathering tools.

TSA-based formats include *differential scanning fluorimetry* (DSF, ThermoFluor[156]) and *isothermal calorimetry* (ITC[157]). DSF is a cheap, HT format suitable as a primary assay ($\geq$1000 compounds/day in optimal conditions) that requires small protein quantities. Thermal unfolding of a protein is altered by its interaction with a fragment, and the temperature variation is measured with a fluorescent reporter dye.[158] Most binding fragments increase the melting temperature, but a decrease may be observed through destabilization of protein folding. The upper detection limit is lower than 1 mM. False positives are frequent, and structural information on target-ligand interactions cannot be obtained.

ITC has a low throughput (tens of compounds/day in optimal conditions), requires larger protein quantities, and is used as a secondary/validation assay. It measures heat consumption (endothermal) or generation (exothermal) caused by protein-fragment interactions, and provides the enthalpic and enthropic parameters of the interaction.[159] Its upper detection limit is also below 1 mM.

MS methods include native electron spray ionization (*native ESI-MS*)[160]; affinity selection (*AS-MS*) that requires a size exclusion chromatography-SEC step to isolate target-fragment complexes[161]; and hydrogen-deuterium exchange (*HDX-MS*) that detects varied H-D exchange rates due to binding and works on membrane-bound proteins.[162] Their throughput varies between ultra-high (AS-MS, using fragment mixtures in primary screens), medium (native MS, ≥100 compounds/day in optimal conditions, primary screening), and low (HDX-MS, tens of compounds/day in optimal conditions). Protein consumption is limited, and the attainable structural information is inversely proportional to their throughput. Their sensitivity is limited (up to 1 mM for native MS, lower than 100 μM for AS- and HDX-MS), as is the occurrence of false positives.

Three biophysical methods were used in most of the reported FBDD-based projects. *Surface plasmon resonance (SPR)*[163] supports a target on a chip in a microfluidic system where solutions of fragments are flowed. The fragment-binding-dependent change of refractive index of the chip surface (*protein-bound SPR*[164]) is measured. Supporting a fragment library on a microarray (*ligand-bound SPR*[165]) and flowing target solutions provides higher throughput (≥10,000 compounds/day in optimal conditions, compared to ≥100 for protein-bound SPR), but increases protein consumption. Both methods were used as primary assays in FBDD. Sensitivity reaches the 1-mM mark, and kinetic and thermodynamic data about the protein-ligand interaction can be obtained. At least one anchoring strategy that does not affect fragment-target binding, or the interaction-relevant conformational freedom of the target, is needed to execute SPR assays avoiding false negatives.

NMR methods[166] include *protein-observed*[167] and *ligand-observed NMR*.[168] The former was the first biophysical method to validate FBDD in drug discovery (*SAR by NMR*[169]). Chemical shift changes in a protein are observed in presence of binding fragments, locating their interaction sites. The dissociation constant can be determined by fragment hit titration, and binding-dependent conformational changes can be observed. Protein-observed NMR requires a labeled target (^{15}N and/or ^{13}C) and works with small proteins (≈50 K MW as an upper limit). Its sensitivity reaches the low mM concentration. False positives and negatives are not frequently observed, but its medium-low throughput (≥100 compounds/day in optimal conditions) makes it unsuitable as a primary assay.

Ligand-observed NMR, and in particular saturation transfer difference (*STD-NMR*[170]), measures changes of the binding fragment (chemical shift, relaxation time, diffusion rates). The throughput is higher, as ligand/fragment mixtures and unlabeled proteins (in larger amounts) can be used, making it suitable as a primary assay. Its sensitivity is similar to protein-observed NMR, but false positives are observed more often and structural information about the binding interaction is limited.

^{19}F-labeled *reporter screen ligand NMR*[171] requires a ^{19}F-labeled weak ligand displaced by binding fragments; *protein-observed* ^{19}F *NMR*[172] incorporates ^{19}F-containing aminoacids in target proteins. As to the former, sensitivity and throughput are suitable for primary screening; as to the latter, there are no MW limits for proteins and its throughput is higher than ^{15}N- and/or ^{13}C-based protein observed NMR. ^{19}F NMR ensures lower protein consumption and faster spectra acquisition for both methods.

X-ray crystallography[166] is the best method for a medicinal chemist, providing a snapshot of the binding fragment in the target-binding site. Crystal complexes are obtained by cocrystallization, or by soaking fragments into preformed protein crystals; the latter does not require the set up of crystallization conditions for each target- fragment pair.[173] Technological advances (i.e., HT experimentation to sample experimental conditions, automated crystallization platforms, miniaturization) enlarged the pool of crystallization-prone proteins, and there is no upper MW limit for a target. Soaking of fragment mixtures (up to 10 different fragments) qualifies X-ray crystallography as a primary assay (≥100 compounds/day in optimal conditions[174]). Crystallographic data collection with syncrothrons increases the throughput up to >12 K crystalline samples evaluated for diffraction quality, and to >4 K X-ray datasets generated in 2011 by a big pharma.[175] The refinement of fragment-binding site interactions can be done manually (biased by the operator, slow) or automatically (prone to misplacements). Automation should be supported by computational energy minimization of a fragment in the binding site, leading to reliable structures.[176]

Nanodrop crystallization plates reduced target consumption, and the sensitivity of X-ray crystallography is extremely high—up to 1 M concentrations.[150] False positives do not happen, as electron density in the target-binding site is specific. Conversely, neither quantitative affinity data about the interaction, nor potency ranking among fragment hits can be gathered. False negatives may happen, especially with soaking, as fragments may not penetrate through the crystal to reach the binding site, and the cosolvent used to solubilize them at high concentrations (usually DMSO) may compromise the crystal quality.[177]

A comparison among described assay formats is summarized in Table 3.1.

There is no universal solution in terms of throughput, sensitivity, avoidance of false positives and negatives, structural information, complexity and cost. If possible in terms of cost and resources, orthogonal techniques (i.e., two or more biophysical assays) are becoming popular.[178] Parallel screening, i.e., the use of two techniques and the selection of fragment hits found in both campaigns, limits the number of false positives. Unfortunately, overlapping hits with different biophysical formats are often few,[179] and parallel screening may lose many binding fragments.[147] A popular choice is *orthogonal sequential screening*, described in a recent paper[180] that illustrates the *modus operandi* in selecting a screening cascade for a successful FBDD example (Fig. 3.21).

A primary HT assay format is selected after validating its suitability for the target, and is used to screen a medium-large fragment collection (1217 fragments in the example); DSF was chosen by Abell (step 1), due to its throughput and low cost. After careful optimization of experimental conditions (step 2), the primary screening was carried out (step 3). Positives from primary screening (57 compounds in the example, ≈5% hit rate) were validated by two ligand-observed NMR methods in the example (step 4a, STD; step 4b, water-ligand

TABLE 3.1 Biophysical Assay Formats: Pros and Cons

Technique	Throughput	Upper Sensitivity	False Positives	False Negatives	Target Consumption	Info Content	Cost
HCS	High	Low	High	High	Medium	Low	Medium
DSF	High	Low	High	High	Low	Low	Low
ITC	Low	Low	Low	Low	High	High	Medium
ESI-MS	High	Medium	Medium	Low	Low	Low	Low
AS-MS	Medium	Low	Low	Low	Low	Medium	Medium
HXD-MS	Low	Low	Low	Low	Medium	High	Medium
SPR-prot. Bound	Medium	Medium	Medium	High	Medium-high	Medium	Medium
SPR-lig. Bound	High	Medium	Medium	High	Low-medium	Medium	Medium
NMR-prot. Observed	Low (^{19}F medium)	High	Low	Low	High (medium ^{19}F)	High	High
NMR-lig. Observed	medium (high ^{19}F)	High	High	Medium	Low-medium	Medium	Medium
X-ray—cocrystalliz.	Low-medium	High	Low	Low	Low-medium	High	High
X-ray—soaking	Medium-high	High	Low	Medium	low-medium	High	Medium

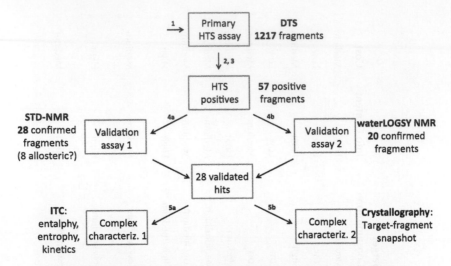

1: selection of a biophysical assay format; 2: optimization of assay conditions; 3: primary screening; 4a: positive fragments' validation, STD-NMR; 4b: positive fragments' validation, waterLOGSY-NMR; 5a: characterization of target-fragment hits' interactions, ITC; 5b: characterization of target-fragment hits' interactions, X-ray crystallography.

FIG. 3.21   Orthogonal sequential screening: an example.

observed gradient spectroscopy - waterLOGSY displacement[181]), to confirm fragment binding with low-protein consumption. The interaction of validated fragment hits (28 fragments confirmed as binders by STD, 20 displaced in waterLOGSY displacement/8 possible allosteric binders in the example, ≈50% hit validation) was structurally characterized to prioritize them for hit-to-lead efforts. ITC (step 5a) and X-ray crystallography (step 5b, Fig. 3.21) were used in the example providing thermodynamic and kinetic data, respectively, and detailed information for further structural optimization efforts.

The other HD pillar—chemical collections—is the focus of the next Chapter. A single example covering in detail biological and chemical aspects of HD is provided in the last Section of Chapter 4.

REFERENCES

1. Andrews, L.; Ralston, S.; Blomme, E.; Barnhart, K. A Snapshot of Biologic Drug Development: Challenges and Opportunities. *Hum. Exp. Toxicol.* **2015**, *34*, 1279–1285.

2. Hughes, J. P.; Rees, S.; Kalindjian, S. B.; Philpott, K. L. Principles of Early Drug Discovery. *Br. J. Pharmacol.* **2011**, *162*, 1239–1249.

3. https://www.acs.org/content/acs/en/education/whatischemistry/landmarks/flemingpenicillin.html.

4. Nirenberg, M.; Leder, P.; Bernfield, M.; Brimacombe, R.; Trupin, J.; Rottman, F.; O'Neal, C. RNA Codewords and Protein Synthesis, VII. On the General Nature of the RNA Code. *Proc. Natl. Acad. Sci. USA* **1965**, *53*, 1161–1168.

5. Schenone, M.; Dančík, V.; Wagner, B. K.; Clemons, P. A. Target Identification and Mechanism of Action in Chemical Biology and Drug Discovery. *Nat. Chem. Biol.* **2013**, *9*, 232–240.

6. Blake, R. Target Validation in Drug Discovery. *Methods Mol. Biol.* **2007**, *356*, 367–377.
7. Lounnas, V.; Ritschel, T.; Kelder, J.; McGuire, R.; Bywater, R. P.; Foloppe, N. Current Progress in Structure-Based Rational Drug Design Marks a New Mindset in Drug Discovery. *Comput. Struct. Biotechnol. J.* **2013**, *5*, e201302011.
8. Sliwoski, G.; Kothiwale, S.; Meiler, J.; Lowe, E. W., Jr. Computational Methods in Drug Discovery. *Pharmacol. Rev.* **2014**, *66*, 334–395.
9. https://www.ncbi.nlm.nih.gov/books/NBK53196/.
10. Noah, J. W. New Developments and Emerging Trends in High-Throughput Screening Methods for Lead Compound Identification. *Int. J. High Throughput Screen.* **2010**, *1*, 141–149.
11. Mayr, L. M.; Bojanic, D. Novel Trends in High-Throughput Screening. *Curr. Opin. Pharmacol.* **2009**, *9*, 580–588.
12. Astle, T. W. Microplate Standardization. *Mol. Online* **1997**, *1*, 106–113.
13. Zhang, J.-H.; Chung, T. D. Y.; Oldenburg, K. R. A Simple Statistical Parameter for Use in Evaluation and Validation of High-Throughput Screening Assays. *J. Biol. Screen.* **1999**, *4*, 67–73.
14. Mayr, L. M.; Fuerst, P. The Future of High-Throughput Screening. *J. Biomol. Screen.* **2008**, *13*, 443–448.
15. Sittampalam, S., Coussens, N. P., Eds. *Assay Guidance Manual*; Eli Lilly and Company and the National Institutes of Health Chemical Genomics Center: Bethesda, MD, 2008.
16. Wingfield, J. Modular Automation for Screening: A Cost/Benefit Analysis. *Drug Discov. World Series* **2009**, 65–70. Summer.
17. Glaxo Wellcome. Robotic Screening. *Nat. Suppl.* **1996**, *384*, 2.
18. Janzen, W. P. Screening Technologies for Small Molecule Discovery: The State of the Art. *Chem. Biol.* **2014**, *21*, 1162–1170.
19. Hasson, S. A.; Inglese, J. Innovation in Academic Chemical Screening: Filling the Gaps in Chemical Biology. *Curr. Opin. Chem. Biol.* **2013**, *17*, 329–338.
20. Klumpp, M.; Boettcher, A.; Becker, D.; Meder, G.; Blank, J.; Leder, L.; Forstner, M.; Ottl, J.; Mayr, L. M. Readout Technologies for Highly Miniaturized Kinase Assays Applicable to High-Throughput Screening in a 1536-Well Format. *J. Biomol. Screen.* **2006**, *11*, 617–633.
21. Tran, T. M.; Lan, F.; Thompson, C. S.; Abate, A. R. From Tubes to Drops: Droplet-Based Microfluidics for Ultrahigh-Throughput Biology. *J. Phys. D: Appl. Phys.* **2013**, *46*, 114004.
22. Du, G.; Fang, Q.; den Toonder, J. M. J. Microfluidics for Cell-Based High Throughput Screening Platforms. A Review. *Analyt. Chim. Acta* **2016**, *903*, 36–50.
23. Kool, Y.; Lingeman, H.; Niessen, W.; Irth, H. High Throughput Screening Methodologies Classified for Major Drug Target Classes According to Target Signaling Pathways. *Comb. Chem. High Throughput Screen.* **2010**, *13*, 548–561.
24. Bevan, P.; Ryder, H.; Shaw, I. Identifying Small-Molecule Lead Compounds: The Screening Approach to Drug Discovery. *Trends Biotechnol.* **1995**, *13*, 115–121.
25. Zhang, Z.; Guan, N.; Li, T.; Mais, D. E.; Wang, M. Quality Control of Cell-Based High-Throughput Drug Screening. *Acta Pharm. Sin. B.* **2012**, *2*, 429–438.
26. Macarron, R.; Hertzberg, R. P. Design and Implementation of High Throughput Screening Assays. *Mol. Biotechnol.* **2011**, *47*, 270–285.
27. Fox, S. High throughput screening: New Strategies, Success Rates, and Use of Enabling Technologies. HighTech Business Decisions Market Report. http://www.hightechdecisions.com/reports.html, 2007.
28. Giacomotto, J.; Segalat, L. High-Throughput Screening and Small Animal Models, Where Are We? *Br. J. Pharmacol.* **2010**, *160*, 204–216.

29. Wagner, B. K. The Resurgence of Phenotypic Screening in Drug Discovery and Development. *Exp. Opin. Drug Discov.* **2016**, *11*, 121–125.
30. Reynolds, C. H., Merz, K. M., Ringe, D., Eds. *Drug Design: Structure- and Ligand-Based Approaches*; 1st ed.; Cambridge University Press: Cambridge, UK, 2010. 270 p.
31. An, W. F.; Tolliday, N. Cell-Based Assays for High-Throughput Screening. *Mol. Biotechnol.* **2010**, *45*, 180–186.
32. Cook, N. D. Scintillation Proximity Assay: A Versatile High-Throughput Screening Technology. *Drug Discov. Today* **1996**, *1*, 287–294.
33. Simeonov, A.; Jadhav, A.; Thomas, C. J.; Wang, Y.; Huang, R.; Southall, N. T.; Shinn, P.; Smith, J.; Austin, C. P.; Auld, D. S.; Inglese, J. Fluorescence Spectroscopic Profiling of Compound Libraries. *J. Med. Chem.* **2008**, *51*, 2363–2371.
34. Schroeder, K. S.; Neagle, B. D. FLIPR: A New Instrument for Accurate, High Throughput Optical Screening. *J. Biomol. Screen.* **1996**, *1*, 75–80.
35. Giuliano, K. A., Haskins, J. R., Eds. *High Content Screening: A Powerful Approach to Systems Cell Biology and Drug Discovery*; Humana Press: Totowa, NJ, 2010. 435 p.
36. Lea, W. A.; Simeonov, A. Fluorescence Polarization Assays in Small Molecule Screening. *Expert Opin. Drug Discov.* **2011**, *6*, 17–32.
37. Boute, N.; Jockers, R.; Issad, T. The Use of Resonance Energy Transfer in High-Throughput Screening: BRET Versus FRET. *Trends Pharm. Sci.* **2002**, 351–354.
38. Degorce, F.; Card, A.; Soh, S.; Trinquet, E.; Knapik, G. P.; Xie, B. HTRF: A Technology Tailored for Drug Discovery: A Review of Theoretical Aspects and Recent Applications. *Curr. Chem. Genomics* **2009**, *2*, 22–32.
39. Eglen, R. M.; Reisine, T.; Roby, P.; Rouleau, N.; Illy, C.; Bossé, R.; Bielefeld, M. The Use of AlphaScreen Technology in HTS: Current Status. *Curr. Chem. Genomics* **2008**, *1*, 2–10.
40. Antony, S.; Marchand, C.; Stephen, A. G.; Thibaut, L.; Agama, K. K.; Fisher, R. J.; Pommier, Y. Novel High-Throughput Electrochemiluminescent Assay for Identification of Human Tyrosyl-DNA Phosphodiesterase (Tdp1) Inhibitors and Characterization of Furamidine (NSC 305831) as an Inhibitor of Tdp1. *Nucleic Acid Res.* **2007**, *35*, 4474–4484.
41. Hashimoto, J.; Watanabe, T.; Seki, T.; Karasawa, S.; Izumikawa, M.; Seki, T.; Iemura, S.; Natsume, T.; Nomura, N.; Goshima, N.; Miyawaki, A.; Takagi, M.; Shin-Ya, K. Novel In Vitro Protein Fragment Complementation Assay Applicable to High-Throughput Screening in a 1536-Well Format. *J. Biomol. Screen.* **2009**, *14*, 970–979.
42. Black, C. B.; Duensing, T. D.; Trinkle, L. S.; Dunlay, R. T. Cell-Based Screening Using High-Throughput Flow Cytometry. *Assay Drug Dev. Technol.* **2011**, *9*, 13–20.
43. Du, Y.; Xu, J.; Fu, H.; Xu, A. S. In *Label-Free Biosensor Technologies in Small Molecule Modulator Discovery. Chemical Genomics*; Fu, H., Ed.; Cambridge University Press: Cambridge, 2012; pp 245–258.
44. Frearson, J. A.; Collie, I. T. HTS and Hit Finding in Academia – From Chemical Genomics to Drug Discovery. *Drug Discov. Today* **2009**, *14*, 1150–1158.
45. Frye, S. V. The Art of the Chemical Probe. *Nat. Chem. Biol.* **2010**, *6*, 159–161.
46. Lahana, R. Who Wants to Be Irrational? *Drug Discov. Today* **2003**, *8*, 655–656.
47. Garnier, J. Rebuilding the R&D Engine in Big Pharma. *Harvard Bus. Rev.* **2008**, *86*, 68–76.
48. Macarron, R.; Banks, M. N.; Bojanic, D.; Burns, D. J.; Cirovic, D. A.; Garyantes, T.; Green, D. V. S.; Hertzberg, R. P.; Janzen, W. P.; Paslay, J. W.; Schoppfer, U.; Sittampalam, G. S. Impact of High Throughput Screening in Biomedical Research. *Nat. Rev. Drug Discov.* **2011**, *10*, 188–195.
49. DiMasi, J. A.; Grabowski, H. G.; Hansen, R. W. Innovation in the Pharmaceutical Industry: New Estimates of R&D Costs. *J. Health Econ.* **2016**, *47*, 20–33.

50. Ekins, S., Ed. *Computer Applications in Pharmaceutical Research and Development*; John Wiley & Sons: Hoboken, NJ, USA, 2006. 844 p.

51. Xia, X. Bioinformatics and Drug Discovery. *Curr. Topics Med. Chem.* **2017**, *17*, 1709–1726.

52. Katsila, T.; Spyroulias, G. A.; Patrinos, G. P.; Matsoukas, M.-T. Computational Approaches In Target Identification And Drug Discovery. *Comput. Struct. Biotechnol. J.* **2016**, *14*, 177–184.

53. Dingemanse, J.; Krause, A. Impact of Pharmacokinetic-Pharmacodynamic Modelling in Early Clinical Drug Development. *Eur. J. Pharm. Sci.* **2017**, *109S*, S53–S58. https://doi.org/10.1016/j.ejps.2017.05.042.

54. Mason, J. S. Introduction to the Volume and Overview of Computer-Assisted Drug Design in the Drug Discovery Process. In *Comprehensive Medicinal Chemistry II*; Taylor, J. B., Triggle, D. J., Eds.; Vol. 4; Elsevier Ltd.: Oxford, UK, 2006; pp 4–11.

55. Ilari, A.; Savino, C. Protein Structure Determination by X-Ray Crystallography. *Methods Mol. Biol.* **2008**, *452*, 63–87.

56. Wuerz, J. M.; Kazemi, S.; Schmidt, E.; Bagaria, A.; Guentert, P. NMR-Based Automated Protein Structure Determination. *Arch. Biochem. Biophys.* **2017**, *628*, 24–32.

57. Carroni, M.; Saibil, H. M. Cryo Electron Microscopy to Determine the Structure of Macromolecular Complexes. *Methods* **2016**, *95*, 78–85.

58. Fersht, A. R. From the First Protein Structures to our Current Knowledge of Protein Folding: Delights and Scepticisms. *Nat. Rev. Mol. Cell. Biol.* **2008**, *9*, 650–654.

59. Lionta, E.; Spyrou, G.; Vassilatis, D. K.; Cournia, Z. Structure-Based Virtual Screening for Drug Discovery: Principles, Applications and Recent Advances. *Curr. Topics Med. Chem.* **2014**, *14*, 1923–1938.

60. Sastry, G. M.; Adzhigirey, M.; Day, T.; Annabhimoju, R.; Sherman, W. Protein and Ligand Preparation: Parameters, Protocols, and Influence on Virtual Screening Enrichments. *J. Comput. Aided. Mol. Des.* **2013**, *27*, 221–234.

61. Xu, D.; Zhang, Y. *Ab initio* Protein Structure Assembly Using Continuous Structure Fragments and Optimized Knowledge-Based Force Field. *Proteins* **2012**, *80*, 1715–1735.

62. Latek, D.; Ekonomiuk, D.; Kolinski, A. Protein Structure Prediction: Combining de Novo Modeling with Sparse Experimental Data. *J. Comput. Chem.* **2007**, *28*, 1668–1676.

63. Alexander, N.; Bortolus, M.; Al-Mestarihi, A.; Mchaourab, H.; Meiler, J. De novo High-Resolution Protein Structure Determination from Sparse Spin Labeling EPR Data. *Structure* **2008**, *16*, 181–195.

64. Xiang, Z. Advances in Homology Protein Structure Modeling. *Curr. Prot. Peptide Sci.* **2006**, *7*, 217–227.

65. Hillisch, A.; Pineda, L. F.; Hilgenfeld, R. Utility of Homology Models in the Drug Discovery Process. *Drug Discov. Today* **2004**, *9*, 659–669.

66. Yang, Y.; Faraggi, E.; Zhao, H.; Zhou, Y. Improving Protein Fold Recognition and Template-Based Modeling by Employing Probabilistic-Based Matching between Predicted one-Dimensional Structural Properties of Query and Corresponding Native Properties of Templates. *Bioinformatics* **2011**, *27*, 2076–2082.

67. Moraes, I.; Evans, G.; Sanchez-Weatherby, J.; Newstead, S.; Shaw Stewart, P. D. Membrane Protein Structure Determination: The Next Generation. *Biochim. Biophys. Acta* **2014**, *1838*, 78–87.

68. Todeschini, R.; Consonni, V. Molecular Descriptors for Cheminformatics; 2 Vol.; Wiley-VCH: Weinheim, Germany, 2009. 1257 p.

69. Cocchi, M.; Menziani, M. C.; De Benedetti, P. G.; Cruciani, G. Theoretical Versus Empirical Molecular Descriptors in Monosubstituted Benzenes. A Chemometric Study. *Chemom. Intell. Lab. Systems* **1992**, *14*, 209–224.

70. Todeschini, R.; Consonni, V. Handbook of Molecular Descriptors, 2nd ed.; Wiley-VCH: Weinheim, Germany, 2008. 667 p.

71. http://www.moleculardescriptors.eu/softwares/softwares.htm.

72. Irwin, J. J.; Shoichet, B. K. Docking Screens for Novel Ligands Conferring New Biology. *J. Med. Chem.* **2016**, *59*, 4103–4120.

73. Huang, S.-Y.; Grinter, S. Z.; Zou, X. Scoring Functions and their Evaluation Methods for Protein–Ligand Docking: Recent Advances and Future Directions. *Phys. Chem. Chem. Phys.* **2010**, *12*, 12899–12908.

74. https://en.wikipedia.org/wiki/List_of_protein-ligand_docking_software.

75. Kitchen, D. B.; Decornez, H.; Furr, J. R.; Bajorath, J. Docking and Scoring in Virtual Screening for Drug Discovery: Methods and Applications. *Nat. Rev. Drug Discov.* **2004**, *3*, 935–949.

76. Li, Y.; Han, L.; Liu, Z.; Wang, R. Comparative Assessment of Scoring Functions on an Updated Benchmark: 2. Evaluation Methods and General Results. *J. Chem. Inf. Model.* **2014**, *54*, 1717–1736.

77. Muegge, I. A Knowledge-Based Scoring Function for Protein-Ligand Interactions: Probing the Reference State. *Perspect. Drug Discov. Des.* **2000**, *20*, 99–114.

78. Danishuddin, M.; Khan, A. U. Structure Based Virtual Screening to Discover Putative Drug Candidates: Necessary Considerations and Successful Case Studies. *Methods* **2015**, *71*, 135–145.

79. Seebeck, B.; Reulecke, I.; Kämper, A.; Rarey, M. Modeling of Metal Interaction Geometries for Protein-Ligand Docking. *Proteins* **2008**, *71*, 1237–1254.

80. Lexa, K. W.; Carlson, H. A. Protein Flexibility in Docking and Surface Mapping. *Q. Rev. Biophys.* **2012**, *45*, 301–343.

81. Sotriffer, C. A.; Kraemer, O.; Klebe, G. Probing Flexibility and "Induced-Fit" Phenomena in Aldose Reductase by Comparative Crystal Structure Analysis and Molecular Dynamics Simulations. *Proteins* **2004**, *56*, 52–66.

82. Ferrari, A. M.; Wei, B. Q.; Costantino, L.; Shoichet, B. K. Soft Docking and Multiple Receptor Conformations in Virtual Screening. *J. Med. Chem.* **2004**, *47*, 5076–5084.

83. Nabuurs, S. B.; Wagener, M.; de Vlieg, J. A Flexible Approach to Induced Fit Docking. *J. Med. Chem.* **2007**, *50*, 6507–6518.

84. Huang, S. Y.; Zou, X. Ensemble Docking of Multiple Protein Structures: Considering Protein Structural Variations in Molecular Docking. *Proteins* **2007**, *66*, 399–421.

85. Craig, I. R.; Essex, J. W.; Spiegel, K. Ensemble Docking into Multiple Crystallographically Derived Protein Structures: An Evaluation Based on the Statistical Analysis of Enrichments. *J. Chem. Inf. Model.* **2010**, *50*, 511–524.

86. Abrams, C. F.; Vanden-Eijnden, E. Large-Scale Conformational Sampling of Proteins Using Temperature-Accelerated Molecular Dynamics. *Proc. Natl. Acad. Sci. USA* **2010**, *107*, 4961–4966.

87. Korb, O.; Olsson, T. S.; Bowden, S. J.; Hall, R. J.; Verdonk, M. L.; Liebeschuetz, J. W.; Cole, J. C. Potential and Limitations of Ensemble Docking. *J. Chem. Inf. Model.* **2012**, *52*, 1262–1274.

88. Clark, A. J.; Tiwary, P.; Borrelli, K.; Feng, S.; Miller, E. B.; Abel, R.; Friesner, R. A.; Berne, B. J. Prediction of Protein–Ligand Binding Poses Via a Combination of Induced Fit Docking and Metadynamics Simulations. *J. Chem. Theory Comput.* **2016**, *12*, 2990–2998.

89. Leenalanda, S. P.; Lindert, S. Computational Methods in Drug Discovery. *Beilstein J. Org. Chem.* **2016**, *12*, 2694–2718.

90. Klabunde, T.; Giegerich, C.; Evers, A. Sequence-Derived Three-Dimensional Pharmacophore Models for g-Protein-Coupled Receptors and Their Application in Virtual Screening. *J. Med. Chem.* **2009**, *52*, 2923–2932.

91. Ertl, P. Cheminformatics Analysis of Organic Substituents: Identification of the Most Common Substituents, Calculation of Substituent Properties, and Automatic Identification of Drug-Like Bioisosteric Groups. *J. Chem. Inf. Comput. Sci.* **2003**, *43*, 374–380.

92. Xie, X.-Q. Exploiting PubChem for virtual screening. *Expert Opin. Drug Discov.* **2010**, *5*, 1205–1220.

93. Sterling, T.; Irwin, J. J. Zinc 15 – Ligand Discovery for Everyone. *J. Chem. Inform. Model.* **2015**, *55*, 2324–2337.

94. Naderi, M.; Alvin, C.; Ding, Y.; Mukhopadhyay, S.; Brylinski, M. A Graph-Based Approach to Construct Target-Focused Libraries for Virtual Screening. *J. Cheminform.* **2016**, *8*, 14.

95. Deanda, F.; Stewart, E. L.; Reno, M. J.; Drewry, D. H. Kinase-Targeted Library Design through the Application of the PharmPrint Methodology. *J. Chem. Inform. Model.* **2008**, *48*, 2395–2403.

96. Zhu, T.; Cao, S.; Su, P.-C.; Patel, R.; Shah, D.; Chokshi, H. B.; Szukala, R.; Johnson, M. E.; Hevener, K. E. Hit Identification and Optimization in Virtual Screening: Practical Recommendations Based Upon a Critical Literature Analysis. *J. Med. Chem.* **2013**, *56*, 6560–6572.

97. Baell, J. B.; Holloway, G. A. New Substructure Filters for Removal of Pan Assay Interference Compounds (PAINS) from Screening Libraries and for their Exclusion in Bioassays. *J. Med. Chem.* **2010**, *53*, 2719–2740.

98. Miteva, M. A.; Violas, S.; Montes, M.; Gomez, D.; Tuffery, P.; Villoutreix, B. O. FAF-Drugs: Free ADME/Tox Filtering of Compound Collections. *Nucl. Acid Res.* **2006**, *34*, W738–W744.

99. Harvey, A. L.; Edrada-Ebel, R. A.; Quinn, R. J. The Re-Emergence of Natural Products for Drug Discovery in the Genomics Era. *Nat. Rev. Drug Discov.* **2015**, *14*, 111–129.

100. Simmons, K. J.; Chopra, I.; Fishwick, C. W. G. Structure-Based Discovery of Antibacterial Drugs. *Nat. Rev. Microbiol.* **2010**, *8*, 501–510.

101. Baell, J. B. Broad Coverage of Commercially Available Lead-Like Screening Space with Fewer Than 350,000 Compounds. *J. Chem. Inf. Model.* **2013**, *53*, 39–55.

102. Gaurav, A.; Gautam, V. Structure-Based three-Dimensional Pharmacophores as an Alternative to Traditional Methodologies. *J. Receptor Ligand Channel Res.* **2014**, (7), 27–38.

103. Cortes-Cabrera, A.; Sanchez Murcia, P. A.; Morreale, A.; Gago, F. Ligand-Based Drug Discovery and Design. *In Silico Drug Discovery and Design*; Cavasotto, C. N., Ed.; CRC Press: Boca Raton, USA, 2016; pp 99–121.

104. Maggiora, G.; Vogt, M.; Stumpfe, D.; Bajorath, J. Molecular Similarity in Medicinal Chemistry. *J. Med. Chem.* **2014**, *57*, 3186–3204.

105. Stumpfe, D.; Bajorath, J. Similarity Searching. *Wiley Interdiscip. Rev.: Comput. Mol. Sci.* **2011**, 260–282.

106. Cereto-Massagué, A.; Ojeda, M. J.; Valls, C.; Mulero, M.; Garcia-Vallvé, S.; Pujadas, G. Molecular Fingerprint Similarity Search in Virtual Screening. *Methods* **2015**, *71*, 58–63.

107. http://www.daylight.com/dayhtml/doc/theory/theory.finger.html.

108. Batista, J.; Bajorath, J. Similarity Searching Using Compound Class-Specific Combinations of Substructures Found in Randomly Generated Molecular Fragment Populations. *ChemMedChem* **2008**, *3*, 67–73.

109. Tovar, A.; Eckhert, H.; Bajorath, J. Comparison of 2D Fingerprint Methods for Multiple-Template Similarity Searching on Compound Activity Classes of Increasing Structural Diversity. *ChemMedChem* **2007**, *2*, 209–217.

110. Carhart, R. E.; Smith, D. H.; Venkataraghavan, R. Atom Pairs as Molecular Features in Structure-Activity Studies: Definition and Applications. *J. Chem. Inf. Comput. Sci.* **1985**, *25*, 64–73.

111. Rogers, D.; Brown, R. D.; Hahn, M. Using Extended Connectivity Fingerprints with Laplacian Modified Bayesian Analysis in High-Throughput Screening Follow-Up. *J. Biomol. Screen.* **2005**, *10*, 682–686.

112. Eckert, H.; Bajorath, J. Design and Evaluation of a Novel Class-Directed 2D Fingerprint to Search for Structurally Diverse Compounds. *J. Chem. Inf. Model.* **2006**, *46*, 2515–2526.

113. Mason, J. S.; Morize, I.; Menard, P. R.; Cheney, D. L.; Hulme, C.; Labaudiniere, R. F. New 4-Point Pharmacophore Method for Molecular Similarity and Diversity Applications: Overview over the Method and Applications, Including a Novel Approach to the Design of Combinatorial Libraries Containing Privileged Substructures. *J. Med. Chem.* **1999**, *42*, 3251–3264.

114. Wermuth, C. G.; Ganellin, C. R.; Lindberg, P.; Mitscher, L. A. Glossary of Terms Used in Medicinal Chemistry (IUPAC Recommendations 1998). *Pure Appl. Chem.* **1998**, *70*, 1129–1143.

115. Acharya, C.; Coop, A.; Polli, J. E.; Mackerell, A. D., Jr. Recent Advances in Ligand-Based Drug Design: Relevance and Utility of the Conformationally Sampled Pharmacophore Approach. *Curr. Comput. Aided Drug Des.* **2011**, *7*, 10–22.

116. Greenidge, P. A.; Carlsson, B.; Bladh, L. G.; Gillner, M. Pharmacophores Incorporating Numerous Excluded Volumes Defined by X-ray Crystallographic Structure in Three-Dimensional Database Searching: Application to the Thyroid Hormone Receptor. *J. Med. Chem.* **1998**, *41*; 2503–2512.

117. Willett, P.; Barnard, J. M.; Downs, G. M. Chemical Similarity Searching. *J. Chem. Inf. Comput. Sci.* **1998**, *38*, 983–996.

118. Bajusz, D.; Rácz, A.; Héberger, K. Why Is Tanimoto Index an Appropriate Choice for Fingerprint-Based Similarity Calculations? *J. Cheminform.* **2015**, *7*, 20.

119. Fligner, M.; Verducci, J.; Blower, P. A Modification of the Jaccard-Tanimoto Similarity Index for Diverse Selection of Chemical Compounds Using Binary Strings. *Technometrics* **2002**, *44*, 110–119.

120. Hert, J.; Willett, P.; Wilton, D. J.; Acklin, P.; Azzaoui, K.; Jacoby, E.; Schuffenhauer, A. Comparison of Fingerprint-Based Methods for Virtual Screening Using Multiple Bioactive Reference Structures. *J. Chem. Inf. Comput. Sci.* **2004**, *44*, 1177–1185.

121. Sun, H.; Tawa, G.; Wallqvist, A. Classification of Scaffold Hopping Approaches. *Drug Discov. Today* **2012**, *17*, 310–324.

122. Nisius, B.; Bajorath, J. Reduction and Recombination of Fingerprints of Different Design Increase Compound Recall and the Structural Diversity of Hits. *Chem. Biol. Drug Des.* **2010**, *75*, 152–160.

123. Willett, P. The Calculation of Molecular Structural Similarity: Principles and Practice. *Mol. Inf.* **2014**, *33*, 403–413.

124. Martin, Y. C.; Kofron, J. L.; Traphagen, L. M. Do Structurally Similar Molecules Have Similar Biological Activity? *J. Med. Chem.* **2002**, *45*, 4350–4358.

125. McGaughey, G. B.; Sheridan, R. P.; Bayly, C. I.; Culberson, J. C.; Kreatsoulas, C.; Lindsley, S.; Maiorov, V.; Truchon, J. F.; Cornell, W. D. Comparison of Topological, Shape, and Docking Methods in Virtual Screening. *J. Chem. Inf. Model.* **2007**, *47*, 1504–1519.

126. Stumpfe, D.; Bajorath, J. Applied Virtual Screening: Strategies, Recommendations, and Caveats. In *Methods and Principles in Medicinal Chemistry. Virtual Screening. Principles, Challenges, and Practical Guidelines*; Sotriffer, C., Ed.; Wiley-VCH: Weinheim, Germany, 2011; pp 73–103.

127. Riniker, S.; Landrum, G. A. Open-Source Platform to Benchmark Fingerprints for Ligand-Based Virtual Screening. *J. Cheminform.* **2013**, *5*, 26.

128. Gillet, V. J. Diversity Selection Algorithms. *Wiley Interdiscip. Rev.: Comput. Mol. Sci.* **2011**, *1*, 580–589.

129. Vuorinen, A.; Schuster, D. Methods for Generating and Applying Pharmacophore Models as Virtual Screening Filters and for Bioactivity Profiling. *Methods* **2015**, *71*, 113–134.

130. Mysinger, M. M.; Carchia, M.; Irwin, J. J.; Shoichet, B. K. Directory of Useful Decoys, Enhanced (DUD-E): Better Ligands and Decoys for Better Benchmarking. *J. Med. Chem.* **2012**, *55*, 6582–6594.

131. Steindl, T. M.; Schuster, D.; Laggner, C.; Langer, T. Parallel Screening: A Novel Concept in Pharmacophore Modeling and Virtual Screening. *J. Chem. Inf. Model.* **2006**, *46*, 2146–2157.

132. Sanders, M. P.; Verhoeven, S.; de Graaf, C.; Roumen, L.; Vroling, B.; Nabuurs, S. B.; de Vlieg, J.; Klomp, J. P. Snooker: A Structure-Based Pharmacophore Generation Tool Applied to Class a GPCRs. *J. Chem. Inf. Model.* **2011**, *51*, 2277–2292.

133. Wolber, G.; Langer, T. LigandScout: 3-D Pharmacophores Derived from Protein-Bound Ligands and their Use as Virtual Screening Filters. *J. Chem. Inf. Model.* **2004**, *45*, 160–169.

134. Kirchmair, J.; Distinto, S.; Markt, P.; Schuster, D.; Spitzer, G. M.; Liedl, K. R.; Wolber, G. *J. Chem. Inf. Model.* **2009**, *49*, 678–692.

135. Sanders, M. P. A.; Barbosa, A. J. M.; Zarzycka, B.; Nicolaes, G. A. F.; Klomp, J. P. G.; de Vlieg, J.; Del Rio, A. Comparative Analysis of Pharmacophore Screening Tools. *J. Chem. Inf. Model.* **2012**, *52*, 1607–1620.

136. Heikamp, K.; Bajorath, J. The Future of Virtual Compound Screening. *Chem. Biol. Drug Des.* **2013**, *81*, 33–40.

137. De Luca, L.; Ferro, S.; Damiano, F. M.; Supuran, C. T.; Vullo, D.; Chimirri, A.; Gitto, R. Structure-Based Screening for the Discovery of New Carbonic Anhydrase VII Inhibitors. *Eur. J. Med. Chem.* **2014**, *71*, 105–111.

138. Legraverend, M.; Tunnah, P.; Noble, M.; Ducrot, P.; Ludwig, O.; Grierson, D. S.; Leost, M.; Meijer, L.; Endicott, J. Cyclin-Dependent Kinase Inhibition by New C-2 Alkynylated Purine Derivatives and Molecular Structure of a CDK2-Inhibitor Complex. *J. Med. Chem.* **2000**, *43*, 1282–1292.

139. Keseru, G. M.; Makara, G. M. Hit Discovery and Hit-to-Lead Approaches. *Drug Discov. Today* **2006**, *11*, 741–748.

140. Duffy, B. C.; Zhu, L.; Decornez, H.; Kitchen, D. B. Early Phase Drug Discovery: Cheminformatics and Computational Techniques in Identifying Lead Series. *Bioorg. Med. Chem.* **2012**, *20*, 5324–5342.

141. Clougherty Genick, C.; Barlier, D.; Monna, D.; Brunner, R.; Bé, C.; Scheufler, C.; Ottl, J. Applications of Biophysics in High-Throughput Screening Hit Validation. *J. Biomol. Screening* **2014**, *19*, 707–714.

142. Lipinski, C. A.; Lombardo, F.; Dominy, B. W.; Feeney, P. J. Experimental and Computational Approaches to Estimate Solubility and Permeability in Drug Discovery and Development Settings. *Adv. Drug Delivery Rev.* **1997**, *23*, 3–25.

143. Fischer, E. Einfluss der Configuration auf die Wirkung der Enzyme. *Berichte Deutsch. Chem. Gesell. Berlin* **1894**, *27*, 2985–2993.

144. Dobson, A. Chemical Space and Biology. *Nature* **2004**, *432*, 824–828.

145. Doak, B. C.; Norton, R. S.; Scanlon, M. J. The Ways and Means of Fragment-Based Drug Design. *Pharmacol. Therap.* **2016**, *167*, 28–37.

146. Keseru, G. M.; Erlanson, D. A.; Ferenczy, G. G.; Hann, M. M.; Murray, C. W.; Pickett, S. D. Design Principles for Fragment Libraries: Maximizing the Value of Learnings from Pharma Fragment-Based Drug Discovery (FBDD) Programs for Use in Academia. *J. Med. Chem.* **2016**, *59*, 8189–8206.

147. Erlanson, D. A.; Fesik, S. W.; Hubbard, R. E.; Jahnke, W.; Jhoti, H. Twenty Years on: The Impact of Fragments on Drug Discovery. *Nat. Rev. Drug Discov.* **2016**, 605–619.

148. Reymond, J.-L. The Chemical Space Project. *Acc. Chem. Res.* **2015**, *48*, 722–730.

149. Boettcher, A.; Ruedisser, S.; Erbel, P.; Vinzenz, D.; Schiering, N.; Hassiepen, U.; Rigollier, P.; Mayr, L. M.; Woelcke, J. Fragment-Based Screening by Biochemical Assays: Systematic Feasibility Studies with Trypsin and MMP12. *J. Biomol. Screening* **2010**, *15*, 1029–1041.

150. Renaud, J. P.; Chung, C.-W.; Danielson, U. H.; Egner, U.; Hennig, M.; Hubbard, R. E.; Nar, H. Biophysics in Drug Discovery: Impact, Challenges and Opportunities. *Nat. Rev. Drug Discov.* **2016**, *15*, 679–698.

151. Genick, C. G.; Barlier, D.; Monna, D.; Brunner, R.; Bé, C.; Scheufler, C.; Ottl, J. Applications of Biophysics in High-Throughput Screening Hit Validation. *J. Biomol. Screening* **2014**, *19*, 7007–7014.

152. Folmer, R. H. A. Integrating Biophysics with HTS-Driven Drug Discovery Projects. *Drug Discov. Today* **2016**, *21*, 491–498.

153. Roddy, T. P.; Horvath, C. R.; Stout, S. J.; Kenney, K. L.; Ho, P.-I.; Zhang, J.-H.; Vickers, C.; Kaushik, V.; Hubbard, B.; Wang, K. Mass Spectrometric Techniques for Label-Free High-Throughput Screening in Drug Discovery. *Anal. Chem.* **2007**, *79*, 8207–8213.

154. Pantoliano, M. W.; Petrella, E. C.; Kwasnoski, J. D.; Lobanov, V. S.; Myslik, J.; Graf, E.; Carver, T.; Asel, E.; Springer, B. A.; Lane, P.; Salemme, F. R. High-Density Miniaturized Thermal Shift Assays as a General Strategy for Drug Discovery. *J. Biomol. Screen.* **2001**, *6*, 429–440.

155. Wu, B.; Barile, E.; De, S. K.; Wei, J.; Purves, A.; Pellecchia, M. High-Throughput Screening by Nuclear Magnetic Resonance (HTS by NMR) for the Identification of PPIs Antagonists. *Curr. Top. Med. Chem.* **2015**, *15*, 2032–2042.

156. McMahon, R. M.; Scanlon, M. J.; Martin, J. L. Interrogating Fragments Using a Protein Thermal Shift Assay. *Austr. J. Chem.* **2013**, *66*, 1502–1506.

157. Ladbury, J. E.; Klebe, G.; Freire, E. Adding Calorimetric Data to Decision Making in Lead Discovery: A Hot Tip. *Nat. Rev. Drug Discov.* **2010**, *9*, 23–27.

158. Niesen, F. H.; Berglund, H.; Vedadi, M. The Use of Differential Scanning Fluorimetry to Detect Ligand Interactions that Promote Protein Stability. *Nat. Protocols* **2007**, *2*, 2212–2221.

159. Ruehmann, E.; Betz, M.; Fricke, M.; Heine, A.; Schaefer, M.; Klebe, G. Thermodynamic Signatures of Fragment Binding: Validation of Direct Versus Displacement ITC Titrations. *Biochim. Biophys. Acta* **2015**, *1850*, 647–656.

160. Vivat Hannah, V.; Atmanene, C.; Zeyer, D.; Van Dorsselaer, A.; Sanglier-Cianferani, S. Native MS: An 'ESI' Way to Support Structure- and Fragment-Based Drug Discovery. *Future Med. Chem.* **2010**, *2*, 35–50.

161. Whitehurst, C. E. Affinity Selection-Mass Spectrometry Screening Techniques for Small Molecule Drug Discovery. *Curr. Opin. Chem. Biol.* **2007**, *11*, 518–526.

162. Chalmers, M. J.; Busby, S. A.; Pascal, B. D.; West, G. M.; Griffin, P. R. Differential Hydrogen/Deuterium Exchange Mass Spectrometry Analysis of Protein–Ligand Interactions. *Expert Rev. Proteom.* **2011**, *8*, 43–59.

163. Chavanieu, A.; Pugniere, M. Developments in SPR Fragment Screening. *Expert Opin. Drug Discov.* **2016**, *11*, 489–499.

164. Huber, W.; Mueller, F. Biomolecular Interaction Analysis in Drug Discovery Using Surface Plasmon Resonance Technology. *Curr. Pharm. Des.* **2006**, *12*, 3999–4021.

165. Neumann, T.; Junker, H. D.; Schmidt, K.; Sekul, R. SPR-Based Fragment Screening: Advantages and Applications. *Curr. Top. Med. Chem.* **2007**, *7*, 1630–1642.

166. Jhoti, H.; Cleasby, A.; Verdonk, M.; Williams, G. Fragment-Based Screening Using X-Ray Crystallography and NMR Spectroscopy. *Curr. Opin. Chem. Biol.* **2007**, *11*, 485–493.

167. Stockman, B. J.; Dalvit, C. NMR Screening Techniques in Drug Discovery and Drug Design. *Prog. Nucl. Mag. Res. Spectrosc.* **2002**, *41*, 183–231.

168. Cala, O.; Guilliere, F.; Krimm, I. NMR-Based Analysis of Protein-Ligand Interactions. *Anal. Bioanal. Chem.* **2014**, *406*, 943–956.

169. Shuker, S. B.; Hajduk, P. J.; Meadows, R. P.; Fesik, S. W. Discovering High-Affinity Ligands for Proteins: SAR by NMR. *Science* **1996**, *274*, 1531–1534.

170. Cala, O.; Krimm, I. Ligand-Orientation Based Fragment Selection in STD NMR Screening. *J. Med. Chem.* **2015**, *58*, 8739–8742.

171. Dalvit, C.; Vulpetti, A. Technical and Practical Aspects of ^{19}F NMR-Based Screening: Toward Sensitive High-Throughput Screening with Rapid Deconvolution. *Magn. Reson. Chem.* **2012**, *50*, 592–597.

172. Gee, C. T.; Arntson, K. E.; Urick, A. K.; Mishra, N. K.; Hawk, L. M. L.; Wisniewski, A. J.; Pomerantz, W. C. K. Protein-Observed ^{19}F-NMR for Fragment Screening, Affinity Quantification and Druggability Assessment. *Nat. Prot.* **2016**, *11*, 1414–1427.

173. Hartshorn, M. J.; Murray, C. W.; Cleasby, A.; Frederickson, M.; Tickle, I. J.; Jhoti, H. Fragment-Based Lead Discovery Using X-Ray Crystallography. *J. Med. Chem.* **2005**, *48*, 403–413.

174. Larsson, A.; Jansson, A.; Åberg, A.; Nordlund, P. Efficiency of Hit Generation and Structural Characterization in Fragment-Based Ligand Discovery. *Curr. Opin. Chem. Biol.* **2011**, *15*, 482–488.

175. Wasserman, S. R.; Koss, J. W.; Sojitra, S. T.; Morisco, L. L.; Burley, S. K. Rapid-Access, High-Throughput Synchrotron Crystallography for Drug Discovery. *Trends Pharm. Sci.* **2012**, *33*, 261–267.

176. Mooij, W. T.; Hartshorn, M. J.; Tickle, I. J.; Sharff, A. J.; Verdonk, M. L.; Jhoti, H. Automated Protein–Ligand Crystallography for Structure Based Drug Design. *Chem. Med. Chem.* **2006**, *1*, 827–838.

177. Winter, A.; Higueruelo, A. P.; Marsh, M.; Sigurdardottir, A.; Pitt, W. R.; Blundell, T. L. Biophysical and Computational Fragment-Based Approaches to Targeting Protein-Protein Interactions: Applications in Structure-Guided Drug Discovery. *Quart. Rev. Biophys.* **2012**, *45*, 383–426.

178. http://practicalfragments.blogspot.it/.

179. Schiebel, J.; Radeva, N.; Köster, H.; Metz, A.; Krotzky, T.; Kuhnert, M.; Diederich, W. E.; Heine, A.; Neumann, L.; Atmanene, C.; Roecklin, D.; Vivat-Hannah, V.; Renaud, J.-P.; Meinecke, R.; Schlinck, N.; Sitte, A.; Popp, F.; Zeeb, M.; Klebe, G. One Question, Multiple Answers: Biochemical and Biophysical Screening Methods Retrieve Deviating Fragment Hit Lists. *ChemMedChem* **2015**, *10*, 1511–1521.

180. Mashalidis, E. H.; Sledz, P.; Lang, S.; Abell, C. A three-Stage Biophysical Screening Cascade for Fragment-Based Drug Discovery. *Nat. Protocols* **2013**, *8*, 2309–2324.

181. Dalvit, C.; Fogliatto, G.; Stewart, A.; Veronesi, M.; Stockman, B. WaterLOGSY as a Method for Primary NMR Screening: Practical Aspects and Range of Applicability. *J. Biomol. NMR* **2001**, *21*, 349–359.

[164] Dalvit, D. C., Fagerness, P. E., Knoerzer, D. Fragment-Based Screening Using NMR. *Concepts in Magnetic Resonance...*

[165] Cala, O., Guillière, F., Krimm, I. NMR-Based Analysis of Protein-Ligand Interactions. *Anal. Bioanal. Chem.* 2014, 406, 943–956.

[166] Sanchez-Pedregal, V. M., Reese, M., Meiler, J., Blommers, M. J. J., Griesinger, C., Carlomagno, T. The INPHARMA Method: Protein-Mediated Interligand NOEs for Pharmacophore Mapping. *Angew. Chem., Int. Ed.* 2005, 44, 4172–4175.

[167] Cala, O., Krimm, I. Ligand-Orientation Based Fragment Selection in STD NMR Screening. *J. Med. Chem.* 2015, 58, 8739–8742.

[168] Dalvit, C., Vulpetti, A. Technical and Practical Aspects of 19F NMR-Based Screening: Toward Sensitive High-Throughput Screening with Rapid Deconvolution. *Magn. Reson. Chem.* 2012, 50, 592–597.

[169] Jordan, J. B., Poppe, L., Xia, X., Cheng, A. C., Sun, Y., Michelsen, K., Eastwood, H., Schnier, P. D., Nixey, T., Zhong, W. Fragment Based Drug Discovery: Practical Implementation Based on 19F NMR Spectroscopy. *J. Med. Chem.* 2012, 55, 678–687.

[170] Jahnke, W., Grotzfeld, R. M., Pellé, X., Strauss, A., Fendrich, G., Cowan-Jacob, S. W., Cotesta, S., Fabbro, D., Furet, P., Mestan, J. Binding or Bending: Distinction of Allosteric Abl Kinase Agonists from Antagonists by an NMR-Based Conformational Assay. *J. Am. Chem. Soc.* 2010, 132, 7043–7048.

[171] Unione, L., Galante, S., Díaz, D., Cañada, F. J., Jiménez-Barbero, J. NMR and Molecular Recognition. The Application of Ligand-Based NMR Methods to Monitor Molecular Interactions. *MedChemComm* 2014, 5, 1280–1289.

[172] Ma, R., Wang, P., Wu, J., Ruan, K. Process of Fragment-Based Lead Discovery—A Perspective from NMR. *Molecules* 2016, 21, 854.

Chapter 4

Step IIIb: The Drug-Like Chemical Diversity Pool: Diverse and Targeted Compound Collections

Assay development and validation, high-throughput screening (HTS), hit confirmation and prioritization in *hit discovery* (*HD*) were covered in the previous chapter (biology-oriented HD, phase 3a); compound collections, privileged scaffolds, focused libraries, druggable/rule of 5-compliant and noncompliant libraries are the core of this chapter (chemistry-oriented HD, phase 3b, Fig. 4.1).

The rational purchase, design and/or synthesis of drug-like chemical diversity (CD) became a pillar of hit discovery in recent years. In reality, drug-like CD was instrumental to discover biologically active principles since the inception of drug discovery. Fig. 4.2 sketches a drug discovery paradigm that is valid now as it was thousands of years ago.

To discover new biologically active compounds, one must have access to drug-like CD (*left*), to a biological assay (*bottom*), and to a read-out/discrimination between actives and inactives (*right*, Fig. 4.2). A round of CD access—activity assay—active compound selection is followed by another, driven by the observations from the former round.

Biologically active pollen grains[1] and psychoactive seeds[2] date back to Neanderthal and Neolithic men, respectively. Active principles for centuries afterwards were extracted from, and administered as natural preparations (mostly from plants); their activity was evaluated via a trial-and-error process, i.e., curing diseased individuals from any affliction.

Natural remedies as drug-like CD, and their administration to humans was the cornerstone of therapy for centuries.[3] Progress was slow and hindered by religious beliefs in diseases as punishments for sins,[4] and limited dissemination of test results.[5] Cultural inputs from Greeks, medical treaties from Arabs, herbal preparations from middle age monasteries, medical schools during the Renaissance in Europe led to a systematic and sound approach to disease treatment.[6] Printing press in the 15th century made all of this known to the world.

Isolation of active principles from natural sources was attempted by alchemists, with limited success on inorganic compounds—organic mixtures were

Chemical Sciences in Early Drug Discovery. https://doi.org/10.1016/B978-0-08-099420-8.00004-3

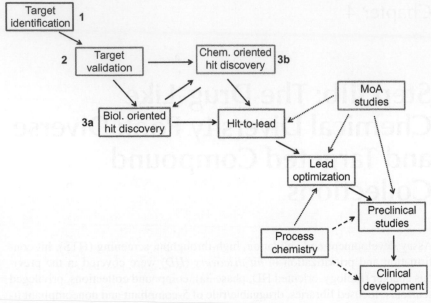

FIG. 4.1 The R&D pharmaceutical process.

The R&D process

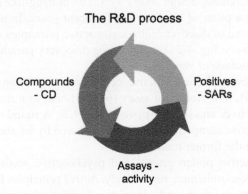

FIG. 4.2 The drug discovery paradigm.

degraded by "hard" distillations or sublimations.[7] Purification techniques in the 19th century enabled compound extraction and characterization. Synthetic compounds were prepared and tested as drugs—chloral hydrate in 1869,[8] antipyrine in 1887,[9] and aspirin in 1897[10] among others.

Scientific and technological improvements, carefully controlled experimentation to find effective drugs, and the demands for a cure against widespread diseases caused around late 19th century the transformation of existing companies (i.e., 1668-founded Merck's pharmacy in Darmstadt, Germany; 1849-founded Pfizer's fine chemicals in Williamsburg, US; 1857-founded Geigy dyestuff production in Basel, Switzerland) into pharmaceutical firms.

Alchemists, monasteries, medical schools, and pharmacies owned small collections of mostly natural remedies. Newly established pharmaceutical companies in late 19th century started to isolate (if working on natural extracts) and/or produce (if synthesizing compounds) biologically active molecules. Each molecule was added to their growing compound collections, to have a reference sample for analytical checks, retesting, additional assays, and so on. Thus, organized chemical collections of drug-like compounds exist since more than a century.

4.1. THE FOUNDATION: DRUG-LIKE CHEMICAL DIVERSITY

The emergence of recombinant DNA technologies and the availability of 96-well microplate-based platforms in early '90s paved the way to in vitro HTS,[11] described in the previous chapter. Thus, large drug-like compound collections since then are a valuable asset in HD, defined by three factors. Their *assembly* is driven by rational (computational similarity, dissimilarity methods) and strategic decisions (collection size, CD sources and targeted subsets). Their *access* depends on acquisition, from vendors and public groups; on (bio)synthesis, using (bio)synthetic strategies and modern technologies; or on a mix of the two. Their *quality* is determined by the thresholds used to select compounds (compound-related filters—purity, physicochemical properties; screening-related filters—reactivity, toxicity, assay interference).

At first, pharmaceutical industries owned medium-large compound collections (up to hundreds of thousand compounds), from decades-long accumulation and storage of synthesized-purchased-isolated compounds. Such collections were not rationally designed, were inaccessible to others, and were used by companies as such in early HTS efforts.[12]

Small molecules in an HTS compound collection are synthesized, biosynthesized, and purchased/accessed, as shown in Fig. 4.3. Computational selection is applied some CD sources.

Compounds were synthesized one by one for centuries by *iterative organic synthesis*. Any synthetic member of early compound collections in large companies was made by iterative synthesis. Unfortunately, the throughput of iterative synthesis was low—a good research lab may produce in a year ≈ 100 final compounds for testing, and HTS craved for more.

Combinatorial chemistry/combichem is the synthesis of large libraries of similar compounds, using semi- or fully automated synthetic protocols in solution or on solid phase (SP)[13] that emerged in the '90s. Early combichem focused on million-membered pool libraries of peptides and nucleotides made by SP chemistry and *mix-and-split synthesis.*[14]

Two limitations of such libraries limited their use in HD. Their sampling of the CD space was limited—polyamide or polyphosphate ester scaffolds with 20 and 4 side chains, respectively, occupy a small CD area; smaller collections of diverse organic molecules better sample the CD space. *Numerosity*, thus, does

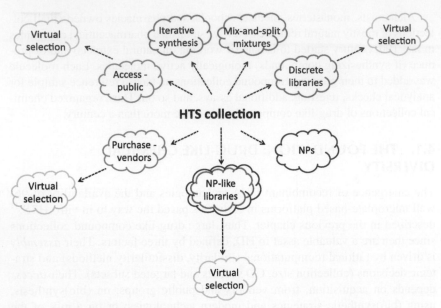

FIG. 4.3 HTS collections: main sources.

not mirror *diversity*: the cartoon in Fig. 4.4 shows how 12 diverse compounds (*right*) sample a simplified, bidimensional CD space better than 50 similar compounds (*left*, Fig. 4.4).

Organic scaffolds replaced peptides and nucleotides in libraries, and were decorated to expand their CD,[15] but the reliability of pools from mix-and-split

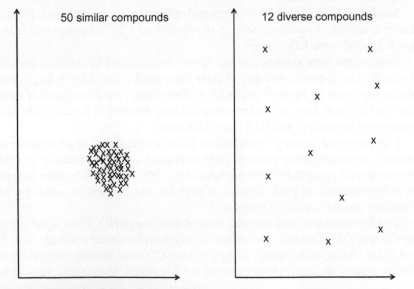

FIG. 4.4 Numerosity (*left*) vs. diversity (*right*).

libraries in HTS was suboptimal. Library representation was an issue—the relative abundance of library members varied (up to 0/complete absence) depending on building block (BB) reactivity. The quality control (QC) of these libraries was limited, and quality issues went undetected. *False negatives*—active compounds undetected due to their concentration/absence in pools—were recurrent. *False positives*—HTS positives, which could not be reconfirmed, due to their inactivity when tested as pure compounds—were found due to unexpected side reactions, and crossreactivity among library members in a pool. An accurate chemical assessment before library synthesis improved the quality of mix-and-split mixtures,[16] but their limitations led to dismissal in HD.

Combichem turned to quality-compliant *discrete libraries*,[17] whose members are fully characterized. Large discrete libraries (up to several tens of thousands individuals) made by parallel synthesis[18] required automation, confining them to combichem-dedicated[19] and big pharma companies. Smaller arrays of discretes (typically hundreds to a few thousand library members) became popular in HD-targeted efforts.[20]

Computational/virtual HTS (*vHTS*) was covered in the previous chapter. The computational evaluation of dissimilarity could drive the selection of diverse small molecules (typically from several hundreds to a few thousands) from a large virtual library built on a core scaffold.[21] Their synthesis would cover most achievable diversity in a molecular scaffold, and would minimize the efforts toward a diverse discrete library.

Activity-based selection of *targeted libraries*[22] for HTS against target classes provided higher hit rates than unbiased larger libraries. Recurrent scaffold-substituent combinations in modulators of protein superfamilies (ligand-based considerations), and/or containing target class-specific structural requirements (structure-based considerations) led to kinase-,[23,24] protease-,[25] G-coupled protein receptors/GPCRs-,[26,27] nuclear receptor-[28] and ion channel-targeted libraries.[29] They covered a large portion of CD space, and diversity-based selection criteria were needed to ensure proper sampling.[30]

Compound collections for HTS benefited from computational drug-likeness filters. The *rule of 5* (*RO5*[31]) was introduced by Lipinski on a drug-like >2200 compound subset from the World Drug Index (WDI). The RO5 defined the optimal range of four properties in orally bioavailable small molecules (Fig. 4.5, *left* clouds).

The *molecular weight* (MW) of a compound is inversely proportional to its passive penetration through a membrane.[32,33] An upper limit of 500 Dalton/Da was set, because only $\approx$11% of the WDI subset had MW $\geq$ 500 Da. The *lipophilicity* of a compound, expressed as the logarithmic partition coefficient among water and n-octanol (cLogP[34]), should be balanced to avoid poor penetration through biological membranes, or poor aqueous solubility. An upper limit of 5 was set for cLogP, because only $\approx$10% of the WDI subset has a cLogP > 5 (a lower limit was not needed, as very hydrophilic compounds are rare). Too many *hydrogen bond donor* (*HBD*) or *acceptor* (*HBA*) groups affect permeability, due to the entropic energy loss to desolvate an HBD-/HBA-rich compound

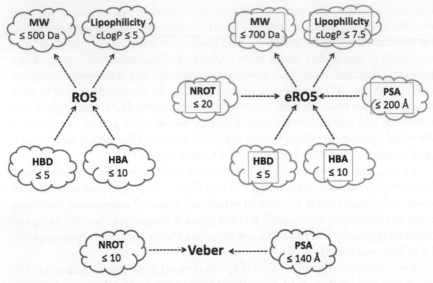

FIG. 4.5 Drug-like CD: RO5, Veber, and eRO5 filters.

in aqueous environments prior to cross nonpolar biological membranes.[35,36] An upper limit of 5 and 10 was set for HBDs and HBAs, respectively, because only ≈8% of the WDI subset had HBD > 5, and only ≈12% had HBA > 10.

Another oral bioavailability filter, based on two properties, was introduced by Veber (Fig. 4.5, *bottom* clouds) on >1100 drug candidates.[37] Molecular rigidity (*number of rotatable bonds/NROT*, upper limit set to 10) and dimensions (*polar surface area/PSA*, upper limit set to 140Å) best correlated with oral bioavailability, claiming performance improvements when compared with RO5.[37]

The RO5, and to a lesser extent the Veber model, was used to filter out nondrug-like compounds either *a priori* (before HTS) on a diversity-based compound collection, or *a posteriori* (after HTS) to prioritize hits from an HTS campaign. Their impact on HD was significant to prioritize putative drug-like/bioavailable hits on oral bioavailability-compliant targets. Soft RO5 versions, substituting numeric thresholds with functions to weigh a violation (i.e., to discriminate compounds with MW = 501 or 800Da), better ranked the drug-likeness of small molecules.[38,39]

Some compounds (i.e., natural products) do not comply with RO5-Veber filters,[40] and some targets (i.e., protein-protein interactions that do not have classical binding pockets[41]) are modulated by non-RO5-compliant molecules. An *extended RO5/eRO5* rule (higher limits for MW, cLogP, NROT, and PSA, Fig. 4.5, *right* clouds) was proposed[42] to modulate unconventional targets with acceptable drug-likeness. Non-RO5-compliant compounds and tough targets are covered in Section 4.3.

Other filters for HTS compound collections are based on *assay interference*. Reactive and/or putative toxic compounds were discarded by experienced medicinal chemists. Nevertheless, the recurrence of several chemotypes as hits in multiple HTS campaigns, and of artifacts in HTS campaigns was observed. A study based on six unrelated HTS campaigns using an AlphaScreen format[43] identified, out of a ≈100K compound collection, several promiscuous compound classes named *pan assay interference compounds (PAINS)*.[44] Assay interference mechanisms were listed by another research group.[45]

PAINS classes contained a higher percentage of multiple hitters (active in at least two of the six HTS campaigns) than inactives (no activity in the six HTS). Some PAINS classes and their interference with HTS assays are depicted in Fig. 4.6.

Rhodamines (235 compounds in the HTS collection, 227% enrichment factor), alkylidene-containing five-membered heterocycles (201/152%), para-quinones (370/265%), catechols (92/97%), 2-hydroxy-phenylhydrazones (479/154%), and 1,2,3-aralkyl pyrroles (118/131%) contained more PAINS compounds.[44] Interference mechanisms include crossreactivity with the HTS

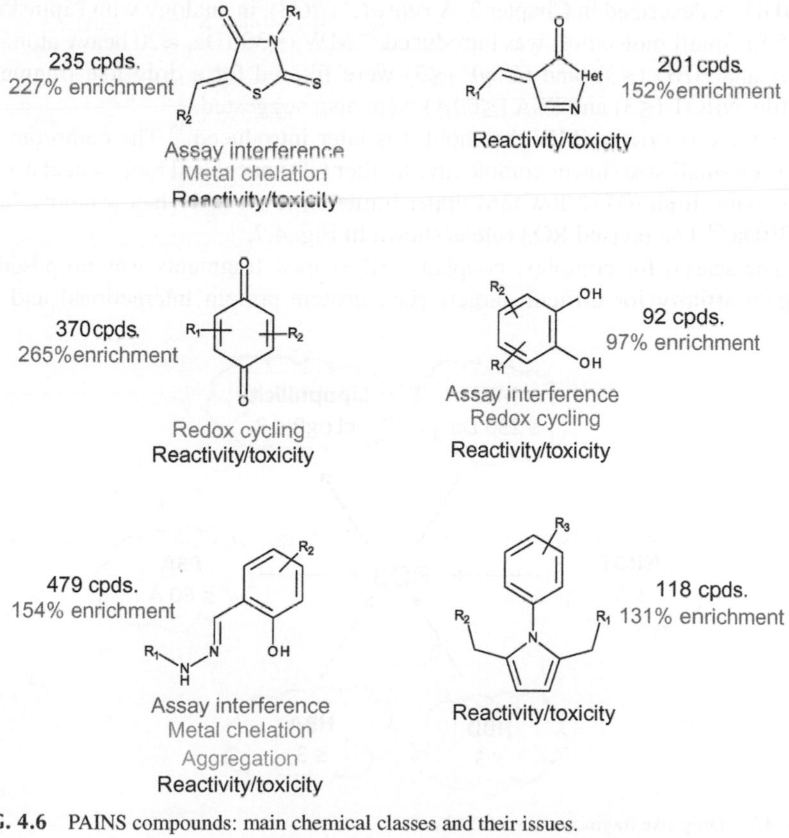

FIG. 4.6 PAINS compounds: main chemical classes and their issues.

detection methods (colored rhodamines and catechols, fluorescent 2-hydroxy-phenylhydrazones); metal chelation (2-hydroxy-phenylhydrazones, rhoda-mines); redox oxidative cycling (catechols, quinones); aspecific reactivity (most PAINS classes); and aggregation propensity (2-hydroxy-phenylhydrazones).

Several papers[46–48] proposed a pre-HTS collection-filtering strategy against PAINS, stating that up to 12% of a compound collection should be discarded to avoid major issues in hit to lead-lead optimization. Some journals requested profiling of hits to avoid assay interference.[49] Conversely, papers arguing against the pre-HTS use of PAINS filters claimed biological specificity for PAINS-flagged compounds,[50,51] including promiscuous/polypharmacology inhibitors.[52] They suggested orthogonal assays to discard verified PAINS compounds from HTS positives, rather than blind PAINS filtering on whole collections.

PAINS filters should be used only to prioritize HTS hits, as they indicate a higher probability of downstream issues in their development. Known drugs (especially from natural sources) belong to PAINS-flagged classes, and the elimination of their structural class from an HTS collection would impact its efficiency as a hit source (especially for difficult targets).

Molecular fragments are the foundation of fragment-based drug discovery (FBDD[53]), described in Chapter 3. A *rule of 3* (*RO3*), in analogy with Lipinski's RO5 for small molecules, was introduced.[54] MW ($\leq$300 Da, $\approx$20 heavy atoms), HBA and HBD ($\leq$3), and cLogP ($\leq$3) were filtered for a drug-like fragment profile. NROT ($\leq$3) and PSA ($\leq$60Å) were also suggested.

A more restrictive MW threshold was later introduced.[55] The compromise between small size (lower complexity, smaller CD space) and biophysical assay sensitivity (high µM to low mM upper limits) was set at 17 heavy atoms, i.e., $\leq$230 Da.[55] The revised RO3 rule is shown in Fig. 4.7.

The search for complex, nonplanar/3D-shaped fragments was proposed[56] to gain affinity for difficult targets (i.e., protein-protein interactions) and to

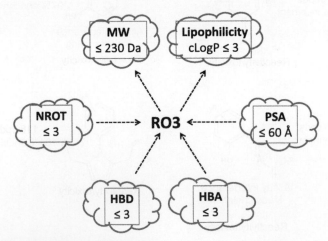

FIG. 4.7 Drug-like fragments: the RO3 filter.

decrease promiscuous binding (attributed to planar fragments). A higher hit rate for 3D-complexes against difficult targets, though, was not confirmed.[57] Rather, a lower overall hit rate should be expected as the average MW for 3D fragments increased and their probability to match a binding site should decrease[58]—see also Fig. 3.19, Chapter 3.

A database of drug-like fragments up to 17 heavy atoms—including stereoisomers—contained $\approx$166.4 billion individuals (GDP-17[59]). Its smaller size (compared to $\approx 10^{28}$ drug-like small molecules[60]) could be sampled by hundreds to tens of thousands fragments (typically 1000–5000[61]).

FBDD collections were assembled from commercial compounds, proprietary collections, and targeted synthesis—big pharma companies[62,63] described their collections and their performance. Their smaller size allowed small labs to manage the rational design,[64] the assembly of an FBDD collection,[65] and its logistics (handling, distribution, QC, etc.).

RO3-compliant collections are preferred in FBDD, but progressable hits from non-RO3-compliant fragments were found.[66] FBDD-driven HD efforts identified recurrent issues to be taken care of.[61,67,68] Frequent issues (top and bottom clouds) and useful features of fragment collections (*lateral* clouds) are shown in Fig. 4.8.

Aqueous solubility is crucial to detect up to low mM affinities in aqueous buffers. Solubility was estimated indirectly by computational methods[69] or directly by HT measurements.[70] Solubility issues jeopardized the reliability of an FBDD collection by affecting its testing at high concentrations (false negatives), or by causing aggregation and precipitation in the assay medium (false positives).

PAINS influenced the quality of FBDD collections more than of small molecules. High concentration testing was *per se* causing compound aggregation

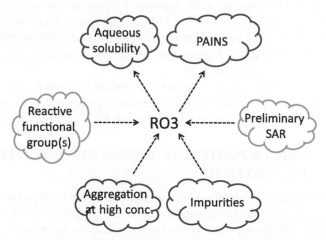

FIG. 4.8   Fragment collections: useful features *(lateral)* and main issues *(top and bottom)*.

even for aggregation-resistant fragments.[71] Reactivity/toxicity testing (Michael acceptors, redox cycling, metal chelation, etc.) was run at high concentrations. Low-level impurities ($\approx$1% in a fragment, i.e., $\approx$10 µM in its 10-mM test solution) could cause assay interference.

Chemically tractable fragments contained at least one reactive function to acquire a SAR.[72] Functional groups were capped (i.e., an amine capped as an acetamide) to mask their nature and make them similar to their analogues (i.e., other amides). Chemical function-dependent capping groups[73] and in silico strategies for SAR acquisition were published.[74,75] Many fragments and fragment collections were reported, but the assessment of synthetic routes to novel, FBDD-compliant fragments is an important goal for medicinal chemists.[76]

Fragment collections should span the fragment CD space and should contain analogues to acquire an initial SAR. A few close analogues in a fragment collection for primary screening, and a secondary analogue-rich fragment set for full SAR acquisition was proposed.[61]

Vendors—suppliers of chemical reagents, or combichem/CD-oriented companies—added many small organic molecules to their catalogues. They offer small-medium size collections for HTS, whose *acquisition* was a cost-effective approach for biology-oriented companies to enable HTS and to discover hits.[77] HD-savvy public and private groups browsed vendor catalogues, and computationally selected and assembled collections of small molecules[77] and fragments[78] from multiple vendors. Even big pharma companies would buy a small subset based on proprietary chemistry sold by a vendor, to enhance their CD space coverage.

Assembled collections from *public sources* are useful for budget-poor academic groups. Initiatives in the US (i.e., the Molecular Libraries Program/MLP at NIH[79,80]) and in Europe (i.e., the European Lead Factory/ELF[81,82]) assembled high-quality HTS collections made by commercially available compounds and libraries, and synthesized libraries from public and private sources. Applications related to R&D/HD projects, after approval by peer review scientific committees, were granted access to HD platforms, including small-molecule- or fragment-based chemical collections, the development of a robust and reliable HTS assay, the run of an HTS campaign, hit identification and validation, and limited hit-to-lead/lead optimization.

The next section describes an example of modern HD, where drug-like hits against a validated target were rationally designed, synthesized, structurally and biologically characterized, and optimized through HTS, vHTS, and FBDD.

4.2. HEAT SHOCK PROTEIN 90 (HSP90): HIT DISCOVERY BY HTS, VHTS, AND FBDD

Misfolded dysfunctional proteins are processed through several paths in cells.[83] Such paths may lead to toxic aggregates, or to refolding/disposal of misfolded protein copies. The former must be antagonized, and the latter must be stimulated to prevent protein aggregation.

The cellular quality control system takes care of misfolded proteins in cells. *Protein chaperones*, such as heat shock protein Hsp70[83] and Hsp90,[84] participate in *protein quality control (PQC)* activities[85,86] that preserve functional proteins in the cell. Chaperones are intrinsically disordered proteins with broad substrate specificity, with high affinity for misfolded proteins and reduced affinity for their folded counterparts.[87]

Chaperone *holdases* hold aggregation-prone, partially misfolded substrates to prevent their aggregation and to stabilize them.[88,89] Chaperone *(un)foldases* bind to misfolded proteins, unfolding and correctly refolding them. This ATP-dependent process leads to properly folded proteins through binding with disordered chaperone regions and localized reorganization.[90,91] Once refolding is complete, (un)foldases and folded client proteins split, due to reduced binding affinity. *Disaggregating chaperones* bind aggregated, misfolded substrates, causing their disaggregation and refolding-detoxification.[91,92] Refolding misfolded proteins reduces the formation of toxic oligomers and provides functionally competent protein copies.[93]

Two major cytoplasmic Hsp90 isoforms, constitutively expressed *Hsp90β* and stress-induced *Hsp90α,*[94] belong to the GHKL ATPase family.[95] They have a >85% sequence homology,[96] and partially different client proteins.[97] The truncated Hsp90N,[97] mitochondrial Hsp75, and endoplasmic-reticulum-located Grp94[98] isoforms are less characterized.

Hsp90 is made by three domains. An N-terminal, ATP-binding domain with ATPase activity (NBD, missing in Hsp90N) precedes a middle segment containing a client protein domain and a catalytic loop to promote ATP hydrolysis (SBD). The C-terminal domain CTD is responsible for Hsp90 dimerization.[04] Hsp90 proteins mostly exist as αα or ββ homodimers, but monomers, heterodimers, and higher oligomers are known.[97]

Binding to ATP forces two C-terminally associated Hsp90 monomers to dimerize through swapping of their N-terminal strand and rotation of a "lid" region in the NBD.[99] The lid closes over the ATP-binding region, and conformational changes bring the middle segment residue Arg380 close to ATP, assisting ATP hydrolysis. ADP-bound and nucleotide-free Hsp90 are open, less defined structures.[88]

Hsp90 functions are largely ATP dependent. Rather than (un)folding and refolding, Hsp90 promotes subtler conformational changes on partially, or even mostly folded proteins.[84] Hsp90 stabilizes bound client proteins, and shows higher binding affinity for mutant/misfolded proteins than for their folded counterparts.[100] Hsp90 often determines the switching between refolding and degradation in a client-dependent manner.[101]

The client list for Hsp90 is updated on the Web,[102] and contains up to 10% of the total protein content of any organism.[103] Clients are structurally unrelated, as is the mechanism of their interaction with Hsp90.[99] Kinases[104] and transcription factors[105] are the most represented client families. The Hsp90 machinery can act as a single chaperone system for a few client proteins,[93] but mostly together with the Hsp70 system.

Hsp90 binds to the heat shock factor 1 (HSF1) and prevents HSF1-dependent heat shock response. Inhibition of Hsp90 leads to HSF1 release, and to stress-activated Hsp70 and Hsp40/J-domain cochaperone induction.[106] The Hsp90 machinery picks up an Hsp70 client protein by binding to the Hop-Hsp70 complex on a different binding site.[107] The ternary Hsp90-Hop-Hsp70 complex then loses Hsp70, binds to Hsp90 cochaperones, refolds, binds the shared Hsp70/Hsp90 cochaperone CHIP[108] on a binding site shared by Hsp70 and Hsp90. Their switch in a CHIP-containing complex drives CHIP-ubiquitinated client proteins toward degradation.[109] In addition to Hsp70-Hsp90 cochaperones Hop and CHIP, Hsp90-specific cochaperones may target it toward client protein subfamilies (cdc37, protein kinases).

Hsp90 is mostly targeted in oncology[84,110] but is relevant against other diseases,[111] such as neurodegeneration.[112] Clinical trials employing Hsp90 inhibitors as anticancer drugs date back to 1998.[113] Since then, tens of clinical candidates have undergone hundreds of oncology trials either as standalone agents, or in combination with cytotoxic agents. Some reviews[110,114,115] describe the status and trends of Hsp90 inhibitors. Here we analyze how HTS, vHTS, and FBDD, virtual and tangible compound or fragment collections assisted the discovery and structural optimization of a few of them.

Most inhibitors bind to the ATP-binding pocket of Hsp90 proteins. Naturally occurring geldanamycin (GM)[116] and radicicol (RC)[117] were identified as low nM NBD binders. Purine-like, ATP-competitive micromolar Hsp90 inhibitors[118] were designed using the ATP-NBD crystal structure. Pyrazoles[119] and isoxazoles[120] were identified as hits in HTS campaigns.

X-rays between the NBD of Hsp90 and GM **1**,[121] RC **2**[122] and ADP **3**[123] were deposited into the Protein Data Bank. They were used at the Memorial Sloan-Kettering Cancer Center (MSK) to rationally design PU3 (**4a**),[112] the first synthetic ATP-competitive Hsp90 inhibitor. Fig. 4.9 depicts molecules **1-4d**, and highlights similarities between PU3 and ADP, PU3 and naturally occurring Hsp90 inhibitors.

PU3 and GM were initially believed to bind similarly to the NBD of Hsp90 (compare panels **A** and **B** in Fig. 4.10).[112] The purine ring of PU3 bound similarly to its counterpart in the ADP-NBD Hsp90 complex, and its trimethoxyphenyl ring was oriented toward the phosphate-binding site.[123] The higher affinity of GM was ascribed to its interaction with the side chain of Lys58.

The abundant structural information on Hsp90 (X-ray[121–124] and NMR studies[125,126]) prompted research groups to focus on synthetic Hsp90 inhibitors. This section deals with the work by Vernalis, either alone or in collaboration with the Institute of Cancer Research (ICR UK) and/or with Novartis. HTS, vHTS campaigns and FBDD-driven efforts, eventually leading to clinical candidates, are described.

Vernalis-ICR paper reported the rational design of PU3 analogues, and the crystallographic characterization of their binding with the NBD of Hsp90α.[127] The PU3-NBD crystals showed relevant differences with the MSK model

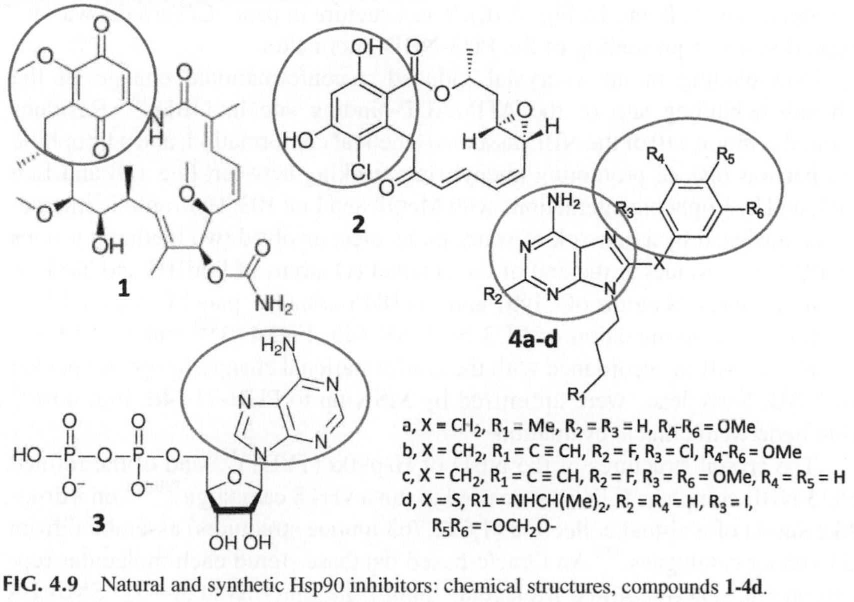

a, X = CH$_2$, R$_1$ – Me, R$_2$ = R$_3$ = H, R$_4$-R$_6$ = OMe
b, X = CH$_2$, R$_1$ = C ≡ CH, R$_2$ = F, R$_3$ = Cl, R$_4$-R$_6$ = OMe
c, X = CH$_2$, R$_1$ = C ≡ CH, R$_2$ = F, R$_3$ = R$_6$ = OMe, R$_4$ = R$_5$ = H
d, X = S, R1 = NHCH(Me)$_2$, R$_2$ = R$_4$ = H, R$_3$ = I,
R$_5$R$_6$ = -OCH$_2$O-

FIG. 4.9   Natural and synthetic Hsp90 inhibitors: chemical structures, compounds **1-4d**.

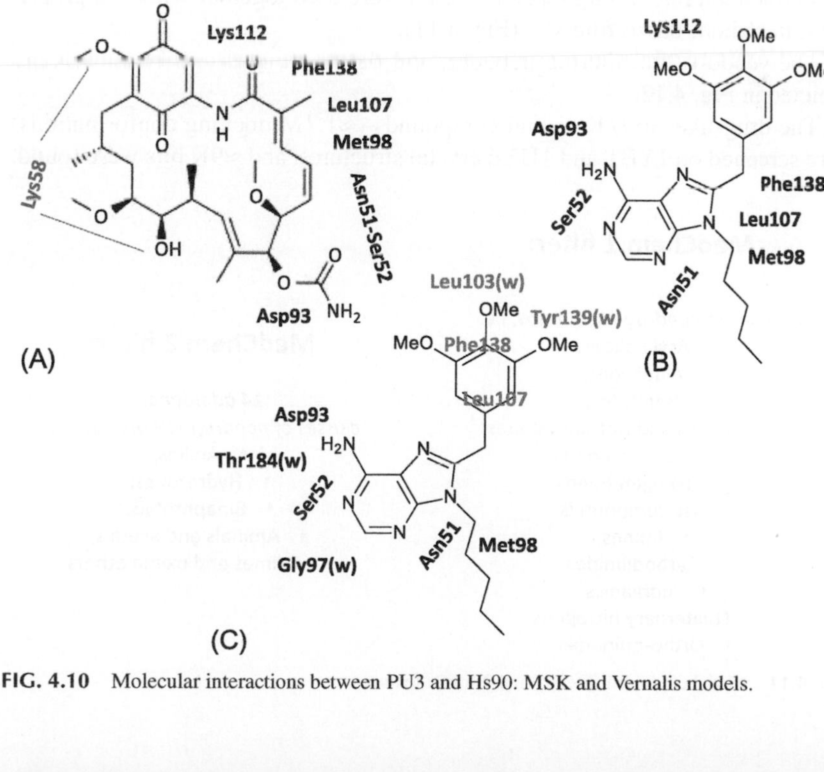

FIG. 4.10   Molecular interactions between PU3 and Hs90: MSK and Vernalis models.

(compare panels **B** and **C**, Fig. 4.10). The structure in panel **C**/Vernalis was accepted as best representing of the PU3-NBD interaction.

PU3 binding in the cocrystal induced a conformational change in the phosphate-binding site of the ATP-/ADP-binding site in NBD.[127] Residues around position 110 of the NBD assumed a helical conformation, and a lipophilic pocket was opened promoting phenyl ring stacking between Phe 138 and Leu 107, and hydrophobic interactions with Met98 and Leu 103. Hydrophilic interactions, mediated by a network of water molecules, involved two methoxy groups of PU3 and residues at the end of the channel (O atoms of Leu103 and Tyr139/ phenolic ethers; N atoms of Gly97 and Thr184/purine Ns, panel **C**, Fig. 4.10).

Structural optimization of PU3 by MSK (**4b**, Fig. 4.9)[128] and ICR (**4c**)[127] showed a SAR in accordance with the conformational change/lipophilic pocket of NBD. Early leads were optimized by MSK up to PUH-71 (**4d**, Fig. 4.9)[129] that underwent clinical evaluation.

The crystal structures of the NBD of Hsp90α (1YET)[121] and of the refined PU3-NBD complex (1UY6) were used to run a vHTS campaign[130,131] on a drug-like subset of a virtual collection (1,622,763 unique structures) assembled from 23 vendor catalogues.[132] An Oracle-based database stored each molecular representation (2D Simplified Molecular Input Line Entry System—2D SMILES descriptors[133]). This collection was analyzed using known and proprietary filters, to select a subset of drug-like molecules.[132] RO5-Lipinski[31] and Veber[37] filters, calculated aqueous solubility,[134] and intestinal permeability[135] ($\geq$10 nm/ sec permeation rate through Caco-2 cells) were used together with two proprietary, medchem-based filters[132] (Fig. 4.11).

The vendors, the filtering network, and the resulting drug-like subsets are depicted in Fig. 4.12.

The drug-like $\approx$607 K virtual compounds ($\approx$1.7 M docking conformations) were screened on 1YET and 1UY6 crystal structures, and $\approx$9K hits were found.

MedChem 1 filter:

50 classes of nondrug-like groups, i.e.
- Acyl halides,
- Aldehydes,
- Isonitriles,
- Isocyanates and isothiocyanates,
- N-halogen bonds
- S-halogen bonds
- Azo compounds
- Imines
- Carbodiimides
- Inorganics
- Quaternary nitrogens
- Ortho-quinones

MedChem 2 filter:

14 additional
classes of nondrug-like groups, i.e.
- Anilins,
- Hydrazones,
- Binaphthtyls,
- Aminals and acetals,
- Oximes and oxime ethers

FIG. 4.11 Medchem-based filters 1 and 2: main features.

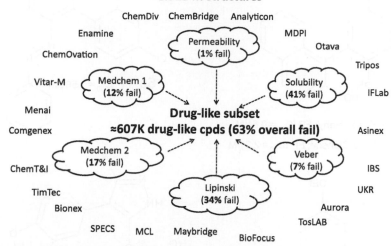

FIG. 4.12 Virtual collection for Hsp90 screening: vendors and filters.

Scoring led to 1000 prioritized hits, with 719 available for purchase. They were tested as inhibitors of the ATPase activity of Hsp90,[136] and 13 compounds (hit rate $\approx$1.8%) showed an $IC_{50} < 100\,\mu$M. Four of them (phenol-naphthol **5a**—two analogues among the vHTS positives; 2-aminopyrimidine **6a**—no analogues; thienopyridine **7a**—one analogue; and resorcinol-pyrazole **8a**—several analogues; Fig. 4.13), showed low μM inhibition in a secondary fluorescence polarization/FP assay.[137]

Hsp90 was targeted by Vernalis also by FBDD, using a chemically tractable and water-soluble (>2 mM in water) collection (1315 fragments divided in four subsets, **SeeDs 1-4**[138]) assembled from commercial vendors. The SeeDs 1 sublibrary came from Aldrich and Maybridge (87,133 starting compounds), as depicted in Fig. 4.14.

An MW filter (110<MW<250 Da., MW<350 Da. if the fragment contained a sulfonamide; step 1, Fig. 4.14) was followed by a structural filter (no 5 linear C sequences, no metals, no reactive groups, step 2). The remaining 7545 fragments were inspected by medicinal chemists ("gut feeling" filter, step 3), selecting 1023 highly soluble and chemically tractable fragments.[138] After purchase, a QC filter (step 4, Fig. 4.14) reduced the SeeDs 1 size to 723 fragments, due to instability in DMSO (200 mM) or test solutions (2 mM in water, $\approx$1% DMSO).

The SeeDs 2 sublibrary was assembled following the process depicted in Fig. 4.15.

The $\approx$1.622M unique structures from 23 vendors (vHTS collection) and the Advance Chemical Directory—ACD[139] database ($\approx$173 K additional structures) were pooled and filtered for MW, as seen for SeeDs 1 (step 1, Fig. 4.15). Fragments were then filtered to avoid unwanted functions ("must not have,"

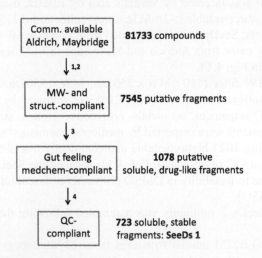

5a, R₁ = 2-thienyl; FP IC₅₀ = 0.6 μM
5b, R₁ = p-Clphenyl; FP IC₅₀ = 2.2 μM

6a FP IC₅₀ = 1.56 μM

7a FP IC₅₀ = 0.9 μM

8a, n = 1, R₁ = H; FP IC₅₀ = 0.8 μM
8b, n = 2, R₁ = Me; FP IC₅₀ = 0.28 μM

FIG. 4.13 Virtual screening on Hsp90: structure of confirmed hits **5-8**.

Comm. available Aldrich, Maybridge	**81733** compounds	

↓ 1,2

MW- and struct.-compliant	**7545** putative fragments	

↓ 3

Gut feeling medchem-compliant	**1078** putative soluble, drug-like fragments	

↓ 4

QC-compliant	**723** soluble, stable fragments: **SeeDs 1**	

1: 110 < MW < 250 (< 350 if sulfonamide-containing); **2**: no linear 5C, metals, reactive groups; **3**: medchem examination (drug-likeness); **4**: QC in 200mM DMSO and 2mM water/1% DMSO.

FIG. 4.14 Fragment collection for NMR screening on Hsp90: sublibrary SeeDs 1.

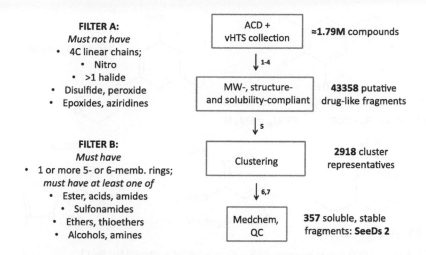

FILTER A:
Must not have
- 4C linear chains;
  - Nitro
  - >1 halide
- Disulfide, peroxide
- Epoxides, aziridines

ACD + vHTS collection   ≈1.79M compounds

↓ 1-4

MW-, structure- and solubility-compliant   43358 putative drug-like fragments

↓ 5

FILTER B:
Must have
- 1 or more 5- or 6-memb. rings;
  must have at least one of
  - Ester, acids, amides
  - Sulfonamides
  - Ethers, thioethers
  - Alcohols, amines

Clustering   2918 cluster representatives

↓ 6,7

Medchem, QC   357 soluble, stable fragments: **SeeDs 2**

1: MW filter as for SeeDs 1; 2: «-» filter A; 3: «+» filter B; 4: predicted ≥2 mM water solubility; 5: clustering, 2D, 3 point pharmacophore-based; 6: medchem filter; 7: QC filter as for SeeDs 1.

FIG. 4.15   Fragment collection for NMR screening on Hsp90: sublibrary SeeDs 2.

filter A, step 2), then to contain desirable moieties ("must have," filter B, step 3). The ≈107K resulting fragments were filtered for predicted solubility (≥2 mM, step 4). Structure clustering of remaining fragments (2D, 3-point pharmacophoric features, step 5) led to 2918 centroids. The 357-membered SeeDs 2 resulted from inspection by medicinal chemists (chemical tractability and cost, step 6), and from QC of purchased fragments (step 7, Fig. 4.15), as seen with SeeDs 1.

The 174-membered, kinase-directed SeeDs 3 sublibrary was assembled by screening with four pharmacophoric functions from the X-ray complex between CDK2 and the small-molecule hymenialdisine[138,140] and filtering (MW/SeeDs1; filter A and B/SeeDs 2; clustering based on Tanimoto Index/similarity; QC/SeeDs 1). The SeeDs 4 sublibrary was obtained by searching the 43,358 soluble, drug-like fragments identified during the assembly of the SeeDs 2 sublibrary (step 4, Fig. 4.15). Novelty of pharmacophores, clustering, inspection by medicinal chemists, and QC of purchased fragments led to a 61 compounds' size for SeeDs 4, i.e., to an overall 1315 size for the SeeDs 1-4 library.[138]

Library members were pooled into mixtures of 12 fragments and screened at 500 μM in presence of the NBD of Hsp90α, using ligand-observed NMR techniques. Saturation transfer difference—STD[141] and water-ligand observed gradient spectroscopy—waterLOGSY displacement[142] were used on SeeDs-1; Carr-Purcell-Meiboom-Gill (CPMG) sequence experiments[143] were added in a second FBDD effort on SeeDs 1-4.[144] In the first FBDD campaign, positive fragments displaced preincubated PU3 (100 μM) from a 10-μM solution of the NBD of Hsp90α.[137] The second FBDD campaign was run at lower concentrations.[144] Six out of ≈60 positive fragments (**9a-13**) are shown in Fig. 4.16.

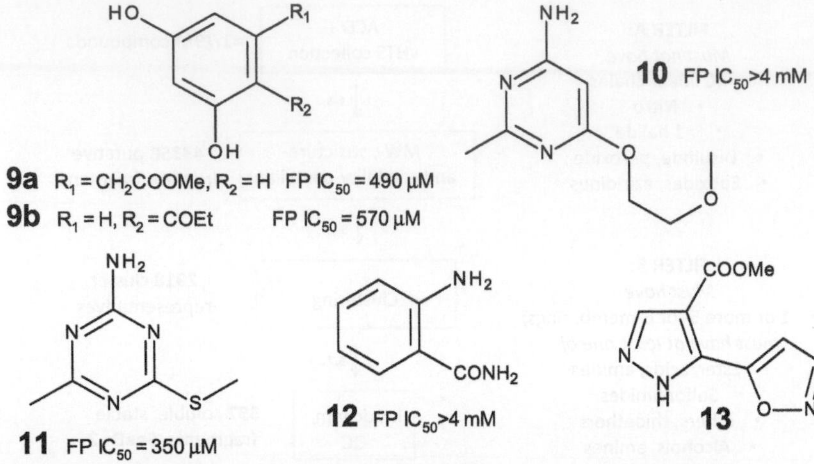

9a $R_1 = CH_2COOMe$, $R_2 = H$ FP $IC_{50} = 490\ \mu M$
9b $R_1 = H$, $R_2 = COEt$ FP $IC_{50} = 570\ \mu M$

10 FP $IC_{50} > 4\ mM$

11 FP $IC_{50} = 350\ \mu M$

12 FP $IC_{50} > 4\ mM$

13

FIG. 4.16   NMR fragment screening on Hsp90: structure of confirmed hits 9a-13.

Similar structural motifs are observed from vHTS (Fig. 4.12) and FBDD-NMR (Fig. 4.16). The resorcinol motif is found in fragments 9a,b and in resorcinol pyrazoles 8; the 2-aminopyrimidine motif is found in fragments 10 and 11 and in the 2-aminopyrimidine 6; the 2-amino carboxamide motif is found in fragment 12 and in thienopyridines 7; and the pyrazole motif is found in fragment 13 and in resorcinol pyrazoles 8.

The ICR group carried out an HTS campaign on a ≈60 K compound collection.[113,145] The HTS assay measured the inhibition of yeast Hsp90 ATPase activity through a malachite green protocol, testing library individuals at 40 μM in 0.4% DMSO.[145] Out of 150 positives (>50% absorbance reduction in the assay), only two resorcinol pyrazoles were confirmed as moderate μM hits[113] in an FP assay format.[137] The potent CCT018159 (8b, Fig. 4.13) was similar to the resorcinol-pyrazole hit 8a from vHTS. It showed low μM cytotoxicity on a panel of cancer cells,[146] and a complex pharmacokinetic—PK profile.[147]

Hit families 5-8 were progressed further at Vernalis, Novartis, and ICR. As to phenol-naphthol 5a, a SAR around it by Vernalis led to compound 5b (Fig. 4.13) and substituted phenylsulfonamides.[148] Their potency was moderate in cellular assays (growth inhibition/GI, HTC116 colon cancer cells). Thus, phenol naphthol-based synthetic inhibitors of Hsp90 were abandoned.

As to aminopyrimidine 6a and thienopyridine 7a, they were pursued with fragments 10-12 by Vernalis and Novartis[149] Fig. 4.17 depicts the structures of modified fragments 14-15 and of related leads 16a-17a,[149] inspired by structural information from X-ray crystals of complexes between putative inhibitors and the NBD domain of Hsp90.

Fragments 14 and 15 showed increased affinity for the Hsp90 NBD compared with fragments 10 (IC_{50} lowered from 4 mM to 535 μM, due to the replacement of a diether side chain with a thioether-amide) and 11 (IC_{50} lowered

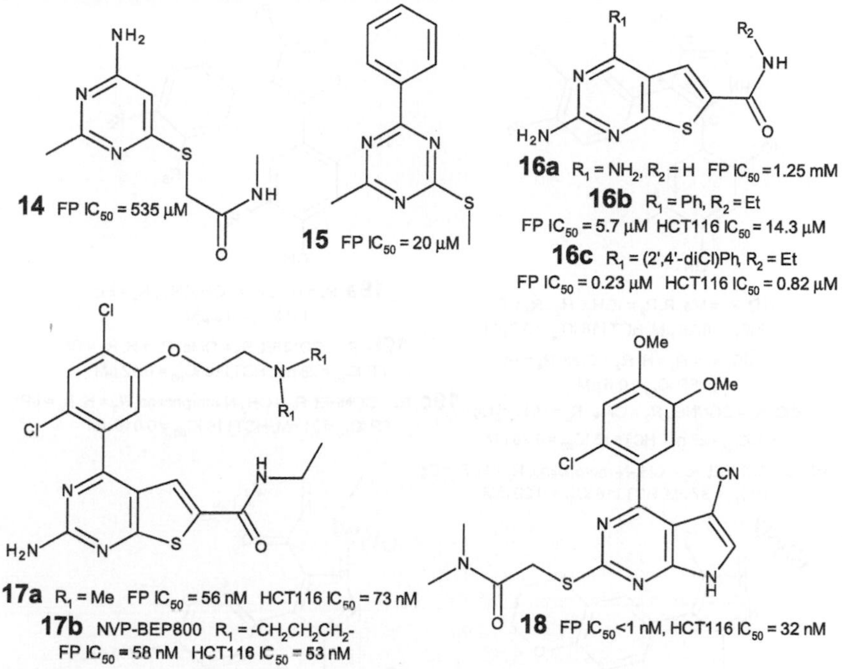

14 FP IC_{50} = 535 μM

15 FP IC_{50} = 20 μM

16a R_1 = NH_2, R_2 = H FP IC_{50} = 1.25 mM
16b R_1 = Ph, R_2 = Et
FP IC_{50} = 5.7 μM HCT116 IC_{50} = 14.3 μM
16c R_1 = (2',4'-diCl)Ph, R_2 = Et
FP IC_{50} = 0.23 μM HCT116 IC_{50} = 0.82 μM

17a R_1 = Me FP IC_{50} = 56 nM HCT116 IC_{50} = 73 nM
17b NVP-BEP800 R_1 = -$CH_2CH_2CH_2$-
FP IC_{50} = 58 nM HCT116 IC_{50} = 53 nM

18 FP IC_{50} < 1 nM, HCT116 IC_{50} = 32 nM

FIG. 4.17 Hit progression at Vernalis: chemical structures, compounds **14-18**.

from 350 to 20 μM, due to the replacement of an amine side chain with a phenyl group). The molecular interactions observed for fragments **10, 11, 14,** and **15,** and for virtual hits **6a** and **7a** inspired the synthesis of an array of thieno-pyridines (**7a**-like, unpublished) and thienopyrimidines (**16a**-like, Fig. 4.17).[149] Both preserved the key interactions with a water molecule network observed with PU3-like inhibitors.[127] The best thienopyrimidines had low μM affinity on Hsp90, and similar potency in cytotoxicity assays, due to a 4-aryl moiety replacing an amine (as seen for fragments **11** and **15**), and to a secondary amide in position 6 (**16b,** Fig. 4.17). o- And p-substitutions on the 4-aryl group with small lipophilic moieties increased potency on Hsp90 and cell penetration, leading to nM inhibitors such as **16c**. Hydrophilic moieties on an affinity-neutral group led to soluble 5'-aryl ethers such as the N,N-diethylamino ethyl ether **17a**.[149] In vivo evaluation of analogues of **17a** led to the selection of preclinical candidate NVP-BEP800 (**17b**).[150] Transplanting 4-aryl substituents and thioether-amide substituents from the thienopyrimidine to the purine/PU3 scaffold led to potent Hsp90 inhibitors (i.e., the preclinical candidate **18,** Fig. 4.17).[151]

A preliminary SAR on resorcinol pyrazoles **5** and related fragments **9** was extracted from an array of pyrazoles, and from a second vHTS (e.g., pyrazole **8c** and isoxazole **19a,** Fig. 4.18[113,131]).

Resorcinol pyrazole **8b** needed an increase in cell-free potency (low nM desired) and cytotoxicity (mid-low nM desired). The X-ray structure of the

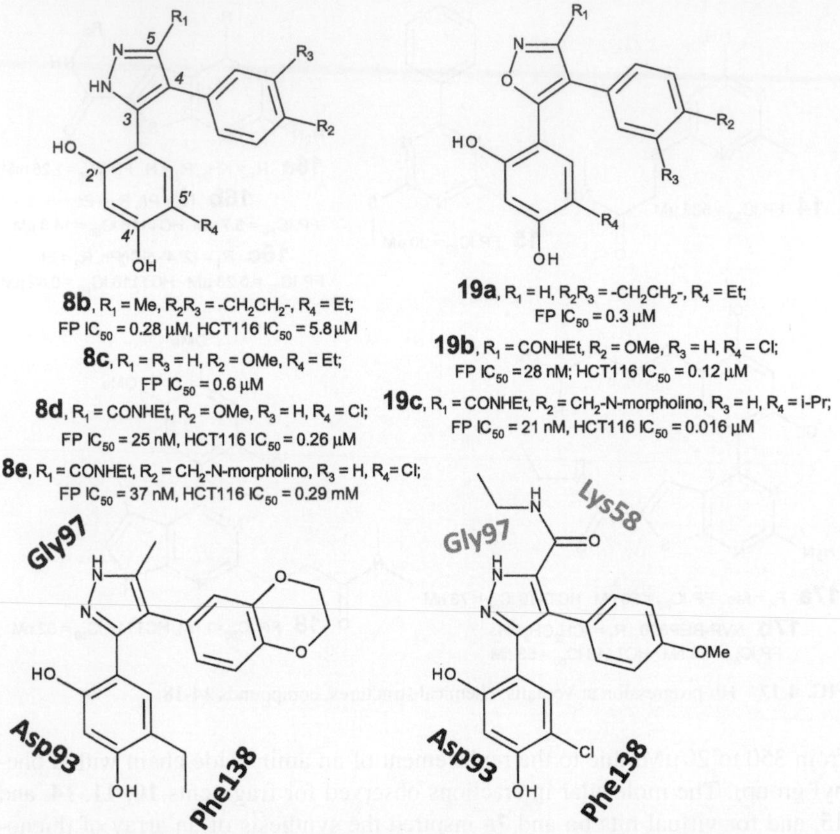

8b, R$_1$ = Me, R$_2$R$_3$ = -CH$_2$CH$_2$-, R$_4$ = Et;
FP IC$_{50}$ = 0.28 µM, HCT116 IC$_{50}$ = 5.8 µM

8c, R$_1$ = R$_3$ = H, R$_2$ = OMe, R$_4$ = Et;
FP IC$_{50}$ = 0.6 µM

8d, R$_1$ = CONHEt, R$_2$ = OMe, R$_3$ = H, R$_4$ = Cl;
FP IC$_{50}$ = 25 nM, HCT116 IC$_{50}$ = 0.26 µM

8e, R$_1$ = CONHEt, R$_2$ = CH$_2$-N-morpholino, R$_3$ = H, R$_4$= Cl;
FP IC$_{50}$ = 37 nM, HCT116 IC$_{50}$ = 0.29 mM

19a, R$_1$ = H, R$_2$R$_3$ = -CH$_2$CH$_2$-, R$_4$ = Et;
FP IC$_{50}$ = 0.3 µM

19b, R$_1$ = CONHEt, R$_2$ = OMe, R$_3$ = H, R$_4$ = Cl;
FP IC$_{50}$ = 28 nM; HCT116 IC$_{50}$ = 0.12 µM

19c, R$_1$ = CONHEt, R$_2$ = CH$_2$-N-morpholino, R$_3$ = H, R$_4$ = i-Pr;
FP IC$_{50}$ = 21 nM, HCT116 IC$_{50}$ = 0.016 µM

FIG. 4.18 Lead progression at Vernalis: chemical structures, resorcinol pyrazoles **8b-e**, and resorcinol isoxazoles **19a-c**.

8b-Hsp90α NBD complex (*bottom left*, Fig. 4.18) drove the structural changes.[152] The resorcinol-like ring in 3 interacted *via* a water network with the Asp93 residue through its 2',4'-OHs, and established a lipophilic interaction with the side chain of Phe138 through its 5'-Et substituent. The bicyclic ring in 4 extended into a solvent-exposed region and did not establish strong interactions. The methyl group in 5 did not interact with the binding site and was positioned close to key residues (Gly97 <4 Å, Lys58 ≈6 Å).

Vernalis and ICR reported ≈10 compounds, introducing an amide bond in position 5 to establish hydrogen bonds with Gly 97 and Lys58, to replace the dioxolane on the resorcinol-like ring with simpler acyclic ethers, and to replace the ethyl group with halogens.[152] Compound VER-49009 (**8d**, Fig. 4.18) showed the desired FP and cellular nM potency. The X-ray structure of the **8d**-Hsp90α NBD complex (*bottom right*, Fig. 4.18) showed the same interactions for the modified resorcinol ring (Asp93 and Phe138 residues), while the amide group established two new HBs with Gly97 through its amidic NH and with Lys58 through its carbonyl group, explaining the increased target affinity.[75] Further attempts to optimize the secondary amide in 5 (N-aryl and N-benzyl substituents[153]), to replace the aryl ring in 4 with cyclic amines (piperidines, piperazines, and morpholines[154]) or to replace the p-methoxy substituent on the 4-aryl ring with solubilizing groups (linear and cyclic amines, i.e., **8e**, Fig. 4.18[155]) could not find a better preclinical candidate than VER-49009/**8d**.

Resorcinol pyrazoles were submitted to scaffold hopping,[156] due to resorcinol isoxazole hits from vHTS (i.e., **19a**, Fig. 4.18). Isoxazole **19b** (VER-50589) overlapped with its pyrazole congener VER-49009/**8d** in a cocrystal structure with NBD, showed similar Hsp90 affinity and potency against HCT116 colon cancer cells,[155] but was more potent against VER-49009-insensitive tumor cell lines.[157] The higher enthalpic binding contribution observed for isoxazole **19b** by isothermal calorimetry/ITC, and a lower dissociation constant/K_d (17 times lower than pyrazole **8d**) could explain its stronger cytotoxicity (nine times more potent in average than pyrazole **8d**) and its cellular accumulation.

The SAR acquired for pyrazoles was confirmed for isoxazoles, through ≈30 analogues (i.e., replacement of 5′-Et and 4-p-OMePh groups[155]). The better drug-likeness of isoxazoles led to bioavailable leads endowed with in vivo activity. Luminespib (NVP-AUY922, **19c**, Fig. 4.18) was selected first for advanced preclinical evaluation,[158] then as a clinical candidate by Novartis. Phase I and/or Phase II studies were performed—the results of some were reported in literature.[159–162]

The use of vHTS, FBDD, and related biophysical techniques on rationally designed/selected chemical diversity rapidly identified hits against Hsp90, and progressed them to preclinical and clinical candidates. Although the development of some of them was put on hold,[163] the merits and the potential of integrated tangible and virtual HD approaches to discover innovative hits are clearly outlined by this example.

4.3. EXPANDING THE DRUG-LIKE SPACE OUT OF THE RO5 BOX

Thinking outside the RO5 box gained recently relevance in a non-RO5-compliant pillar of drug discovery; in a rediscovered alternative source of drug-like CD; and in HT technologies at the CD-biology interface.

Natural products (NPs)[164] were discovered as modulators of using disease-connected phenotypic assays before the elucidation of the genetic code and of molecular targets. Traditional medicine[165] is based upon testing natural extracts (the collection) on diseased humans (the model). The discovery of penicillin in 1929[166] was a meaningful (if serendipitous) assay (killing of pathogens left on a Petri dish) where crude remedies (whole cells of antibiotic-producing moulds) elicited an unexpected biological activity. Extraction, purification, and structural identification of penicillin G from *Penicillum notatum* took more than

10 years and a skilled multidisciplinary team at Oxford,[167,168] due to science and technology limitations.

HTS campaigns on crude or partially purified natural extracts are still run.[169] Progress in taxonomy, molecular and cell biology, extraction and purification protocols, and analytical techniques allowed the assembly of collections of characterized producers/micro-organisms, and of purified natural extracts.[170,171] HTS campaigns identified NPs used as drugs as such, or after semisynthetic modifications. Vancomycin, taxol, and rapamycin (*top*, Fig. 4.19) are NP drugs; aspirin, heroin, and simvastatin (*bottom*, Fig. 4.19) are semisynthetic NP derivatives (from salicylic acid, morphine, and lovastatin, respectively).

Combichem improved the throughput of medicinal chemistry, while NPs represented an unpredictable source of active compounds –no one could have designed the tetracyclic taxol core as a potent microtubule binder before isolating and identifying taxol from *Taxus brevifolia*. Thus, *NP-based libraries*[172,173] became popular.

If an advanced biosynthetic intermediate could be obtained in multigram amounts from an NP producer, its combinatorial *decoration* led to a library of NP analogues.[174] The taxol library **L1** was made by decoration of the protected taxol analogue **20**—obtained from the biosynthetic intermediate baccatin III—with carboxylic acids[175] (Fig. 4.20, *top*).

Sometimes NP scaffolds were synthesized ex novo, as the sarcodictyin library **L2** made by decoration of the synthetic scaffold **21** (obtained in 22 steps from commercial (+)-carvone) with carboxylic acids, isocyanates, alcohols, and amines[176] (Fig. 4.20, *bottom*).

Function-oriented synthesis (*FOS*[177]) fragmented NP scaffolds, identified structural parts responsible for their biological activity, and synthesized simplified active NP analogues. Recurrent structural motifs in NPs active against multiple targets led to synthetic *NP-inspired libraries* such as the large 2,2-dimethylbenzopyran library **L3** from selenium linker-supported intermediates **22**[178,179] (four of ten **L3** sublibraries shown in Fig. 4.21).

A higher content of stereocenters and of O atoms, and a lower number of N atoms characterize NPs vs. synthetic compounds.[180] *Diversity-oriented synthesis* (*DOS*[181]) targeted medium-large libraries built upon complex, stereocenter-rich, polycyclic NP-like scaffolds. The racemic scaffold **23** was converted into 13 diverse polycyclic, stereo-rich scaffolds (i.e., **24-26**, Fig. 4.22) through DOS cascades.[182] DOS was used to synthesize many other NP-like libraries.[183]

Biology-oriented synthesis (*BIOS*[184]) synthesized small-medium NP-like libraries containing 2–4 rings based on biology-targeted scaffolds. Their bioactivity content made them cost-effective additions to compound collections. 3-Substituted 4-benzopyranones **27** were key intermediates reacted with mono- and bisnucleophiles through branching reaction cascades to generate eight decorated molecular frameworks (i.e., **28-30**, Fig. 4.23).[185] Other BIOS-driven efforts in NP-like library synthesis were reported.[186]

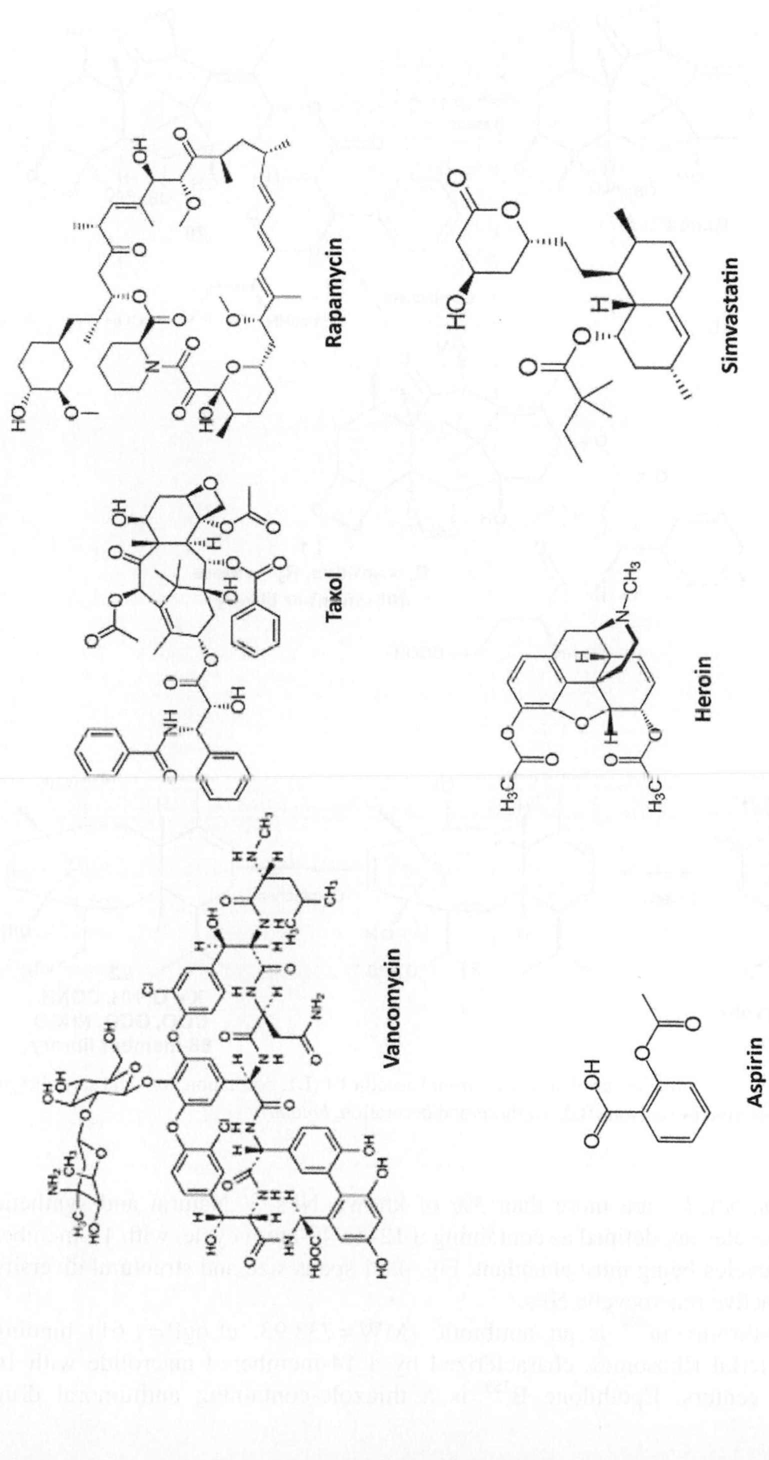

FIG. 4.19 Selected natural products (NP, *top*) and semisynthetic NP derivatives (*bottom*).

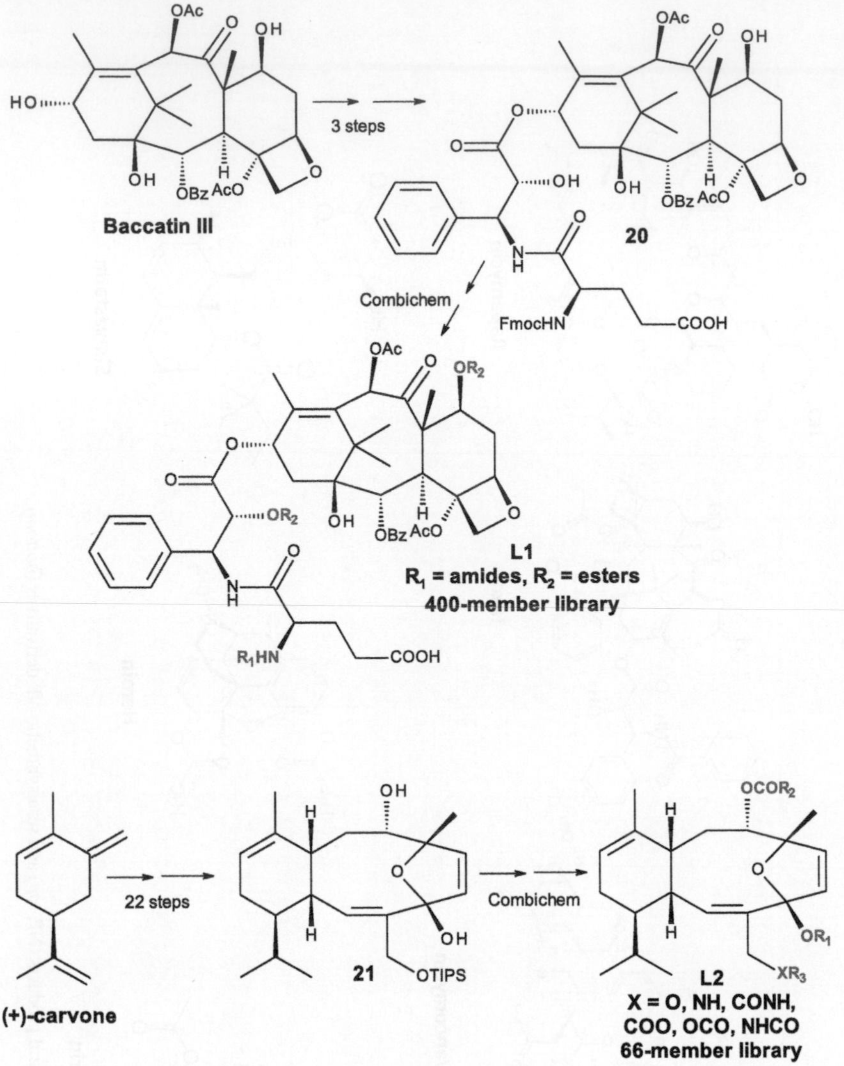

FIG. 4.20 NP libraries: taxol analogues from baccatin III (**L1**, decoration, *top*) and sarcodictyin analogues from (+) carvone (**L2**, synthesis and decoration, *bottom*).

Macrocycles are more than 3% of known NPs.[187] Natural and synthetic macrocycles are defined as containing a 12- to 40-atom cycle, with 14-member macrocycles being most abundant. Fig. 4.24 spans size and structural diversity of bioactive macrocyclic NPs.

Erythromycin[188] is an antibiotic (MW = 733.93, cLogP = 1.61) binding to bacterial ribosomes, characterized by a 14-membered macrolide with 10 chiral centers. Epothilone B[189] is a thiazole-containing antitumoral drug

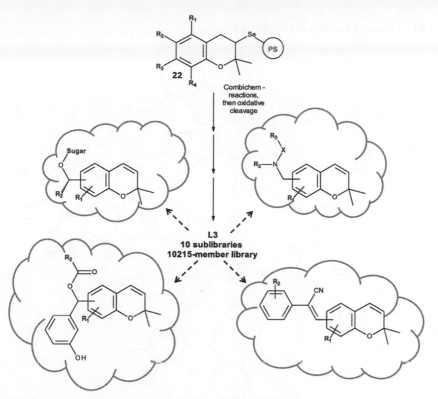

FIG. 4.21 NP libraries: fully synthetic 2,2-dimethylbenzopyran-inspired libraries **L3**.

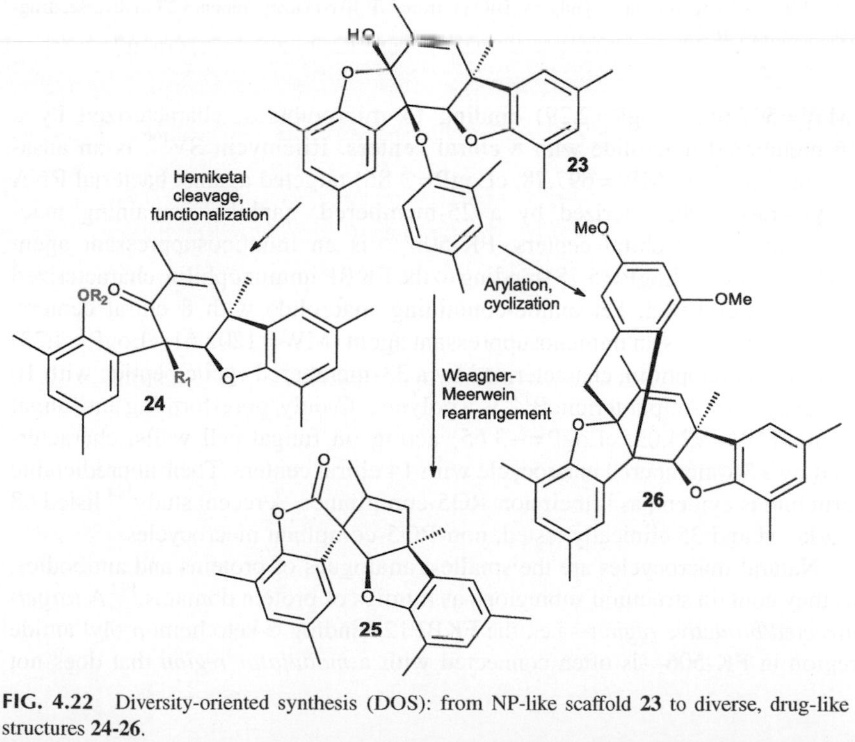

FIG. 4.22 Diversity-oriented synthesis (DOS): from NP-like scaffold **23** to diverse, drug-like structures **24-26**.

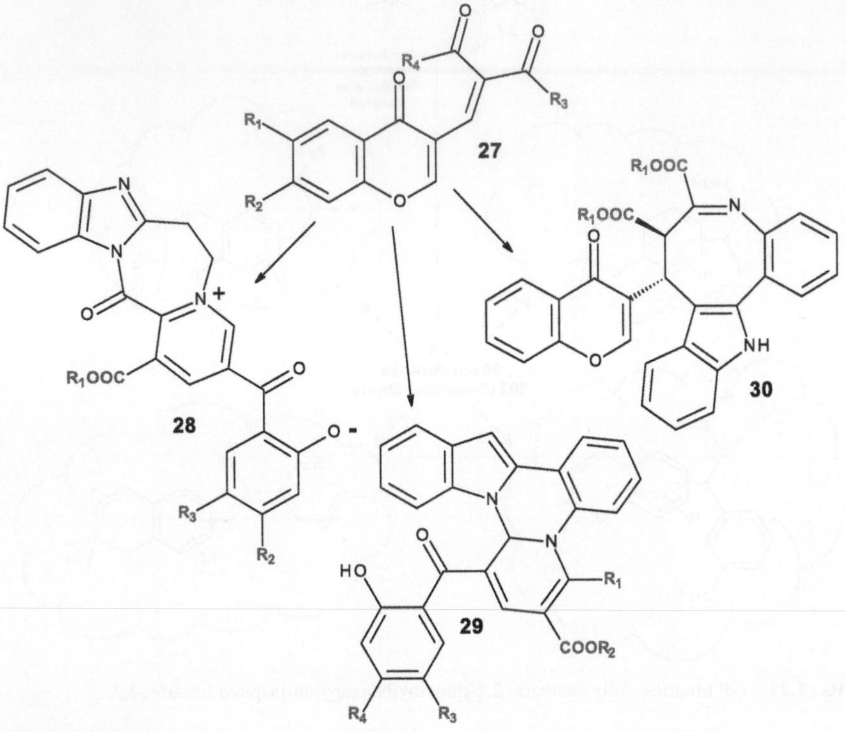

FIG. 4.23 Biology-oriented synthesis (BIOS): from NP-like benzopyranones **27** to diverse, drug-like structures **28-30**.

(MW = 507.68, cLogP = 2.29) binding to microtubules, characterized by a 16-membered macrolide with 8 chiral centers. Rifamycin SV[190] is an ansa-mycin antibiotic (MW = 697.78, cLogP = 3.86) targeted against bacterial RNA polymerases, characterized by a 25-membered, naphthol-containing mac-rolactam with 9 chiral centers. FK-506[191] is an immunosuppressant agent (MW = 804.02, cLogP = 5.78) binding to the FKBP immunophilin, characterized by a 26-membered, ketoamide-containing macrolide with 8 chiral centers. Cyclosporine[192] is an immunosuppressant agent (MW = 1202.61, cLogP = 4.23) binding to ciclophilin, characterized by a 33-membered cyclic peptide with 10 chiral centers. Amphotericin B[193] is a polyene, C-only, pore-forming antifungal agent (MW = 924.09, cLogP = −3.65) acting on fungal cell walls, character-ized by a 38-membered macrocycle with 14 chiral centers. Their unpredictable structure is evident, as is their non-RO5-compliance. A recent study[194] listed 68 marketed and 35 clinically tested, non-RO5-compliant macrocycles.

Natural macrocycles are the smallest analogues of proteins and antibodies, as they contain structural subregions as mimics of protein domains.[195] A *target-directed/bioactive region*—i.e., the FKBP12-binding α-keto homoprolyl amide region in FK-506—is often connected with a *modulator region* that does not

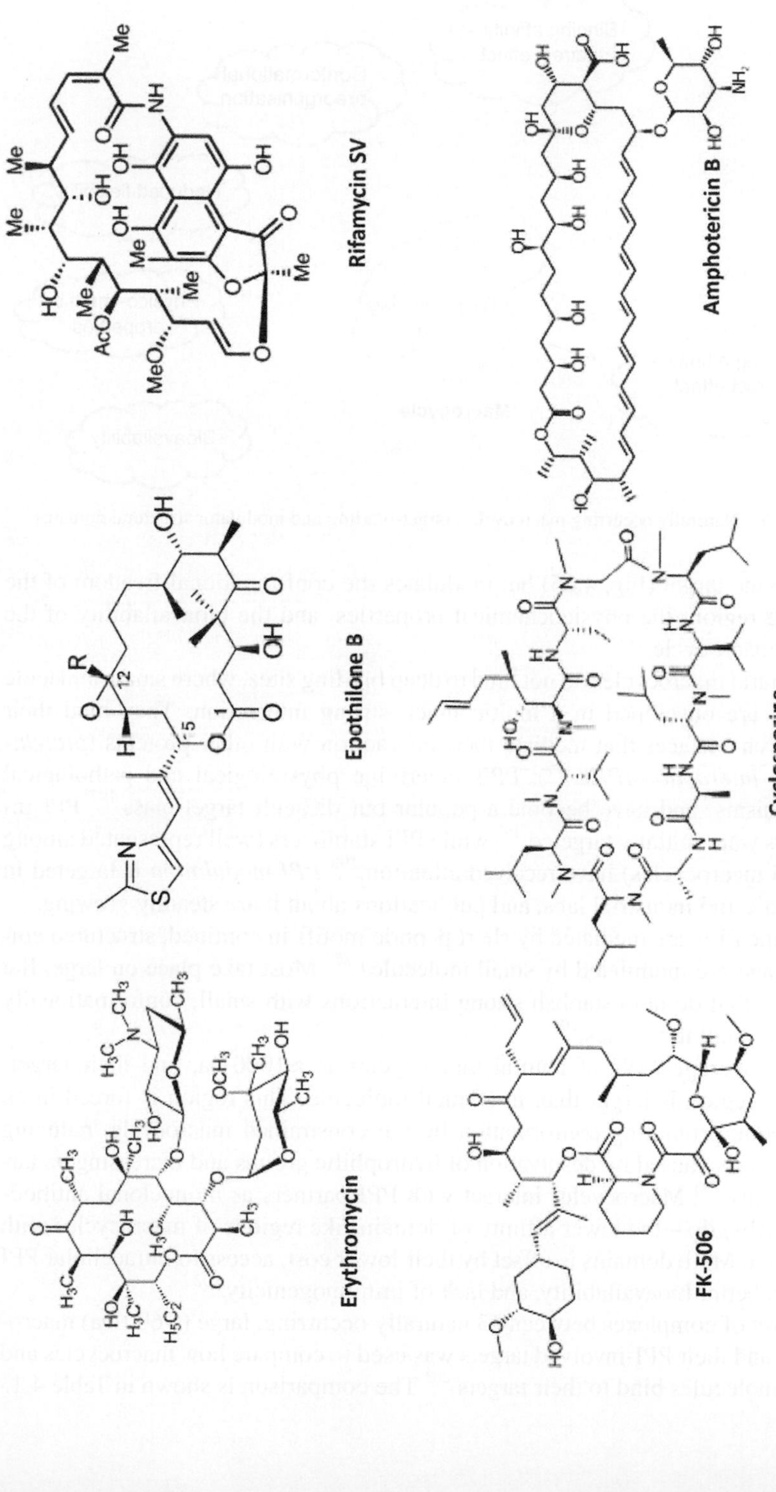

FIG. 4.24 Naturally occurring macrocycles: variations in size and structure.

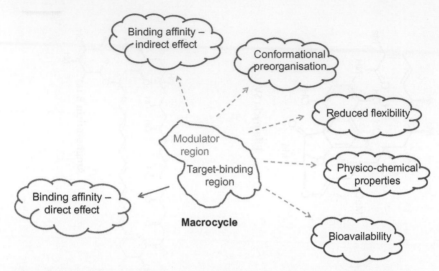

FIG. 4.25 Naturally occurring macrocycles: target-binding and modulator structural domains.

bind to the target (Fig. 4.25) but modulates the conformational freedom of the binding region, the physicochemical properties, and the bioavailability of the whole macrocycle.[195]

Natural macrocycles do not bind to deep binding sites, where small-molecule ligands are enveloped in a multicontact, strong interaction. They bind their targets on surfaces that mediate their interaction with other proteins (*protein-protein interactions/PPIs*[196]). PPIs determine physiological and pathological mechanisms, and have become a popular but difficult target class.[197] PPI inhibitors were initially targeted,[198] while PPI stabilizers (well represented among natural macrocycles) have received attention.[199] *PPI modulation* is targeted in academic and industrial labs, and publications about it are steadily growing.

Some PPIs are mediated by short peptide motifs in confined, structured epitopes, and are modulated by small molecules.[200] Most take place on large, flat surfaces that do not establish strong interactions with small, conformationally flexible small molecules.[201]

The average MW of natural macrocycles is ≈1000 Da, and their target-binding region is larger than in a small molecule. This region is forced in an interaction-promoting conformation by the constrained macrocycle, reducing entropy loss caused by desolvation of hydrophilic groups and increasing its target affinity.[195] Macrocycles interact with PPI partners as monoclonal antibodies (MAbs) do—the lower affinity of domain-like regions of macrocycles with respect to MAb domains is offset by their lower cost, access to intracellular PPI targets, better bioavailability, and lack of immunogenicity.[202]

A set of complexes between 13 naturally occurring, large (>600 Da) macrocycles and their PPI-involved targets was used to compare how macrocycles and small molecules bind to their targets.[203] The comparison is shown in Table 4.1.

TABLE 4.1 Binding to Targets: Differences Among Small Molecules and Macrocycles

Binding Site Features/ Compounds	Small-Molecule-Binding Targets	Large Macrocycle-Binding Targets	Small Macrocycle-Binding Targets
Avg. longest dimension	14.3 Å	16.9 Å	14.4 Å
Avg. number of hot spots	8.2	7.9	7.9
Avg. occupied hot spots	3.6	5.2	5.0
Avg. hot spot separation	8.1 Å	9.3 Å	8.3 Å
Avg separation of two strongest hot spots	7.5 Å	10.6 Å	7.8 Å

Large macrocycle-binding sites (mostly PPI-involved) were ≈15% bigger than classical binding pockets (lane 1, Table 4.1). They showed less hot spots (surfaces with suitable shape and physicochemical properties to establish strong interactions with binders[204]) than smaller but deeper small-molecule-binding sites (lane 2). Conversely, more hot spots were occupied by macrocycles than by small molecules (lane 3), due to the easier adaptation of preorganized macrocycles to larger surfaces. The average distance between hot spots was slightly larger (lane 4), but was much larger among the two strongest hot spots in macrocycle-binding targets (lane 5), as expected for flat, extended PPI surfaces.[203] Smaller macrocycles (400 < MW < 600, right column, Table 4.1) bound to classical binding pockets (lanes 1, 4, and 5), but retained the macrocycle- binding shape to occupy a larger number of hot spots (lane 3).[203]

A linear small molecule is different from a constrained macrocycle binding to similar target-binding sequences in terms of structural and physicochemical properties, but both can be active and bioavailable. The former complies with RO5/Veber filters (first column, Table 4.2); the latter respects alternative criteria,[203] depending on its route of administration (orally available macrocycles, second column; parenterally available macrocycles, third column, Table 4.2).

The comparison of >1700 small-molecule drugs (>1000 orally bioavailable) showed a ≈ 1:3 ratio between N and O vs. C atoms that led to a cLogP ≈ 2.[205] The MW <500 Da prompted other RO5-Veber indicators to fall into the drug-like limits. The ideal ≈1:3 O,N:C ratio should be kept in macrocycles with an MW ≈ 1000 Da to show a suitable cLogP. This O,N:C ratio and MW, though, would double HBAs, HBDs, and PSA values, increasing the entropic penalty needed to desolvate the macrocycle and crossing a biological membrane by passive permeation. Conversely, halving the O,N:C ratio to ≈1:6 would realign HBAs, HBDs, and PSA values, but would increase the cLogP and decrease bioavailability due to poor aqueous solubility.

TABLE 4.2 Properties of In Vivo Active Macrocycles and Small Molecules

Compounds/*Properties*	Oral Small Molecules	Oral Macrocycles	Parenteral Macrocycles
MW	≤500 Da	600–1200 Da	600–1300 Da
cLogP	≤5	−2 to 6	−7 to 2
HBD	≤5	≤12	≤12
HBA	≤10	12–16	9–20
NROT	≤10	≤15	≤17
PSA	≤140 Å²	180–320 Å²	150–500 Å²

Bioavailable macrocycles with an MW ≫ 500 Da did not have a reduced O,N:C ratio, but avoided the energetic penalty/reduced lipophilicity expected with high HBAs, HBDs, and PSA values. A *chameleon effect*[208] (Fig. 4.26) was suggested to explain their bioavailability.

In aqueous environments (*left*, Fig. 4.26), the macrocycle adopted an *exposed/open conformation*, where HBAs (*lighter* dots) and HBDs (darker dots) establish polar solvent interactions. Approaching a biological membrane, the macrocycle adopted an energetically favored *closed conformation* (*right*, Fig. 4.26), where most polar groups interacted intramolecularly. The exposed HBAs and HBDs, and the PSA value decreased, and the macrocycle permeated the nonpolar phase to cross it and access the inner space (i.e., its PPI target(s) in the cytoplasm).[206] Pore formation in membranes,[195] chelation of divalent

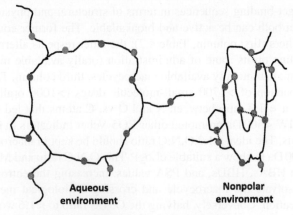

Aqueous environment   **Nonpolar environment**

Red dots: HBDs; Blue dots: HBAs

FIG. 4.26 Macrocycles and bioavailability: the chameleon effect.

cations,[207] and binding to membrane phospholipids[208] could contribute to the higher bioavailability of natural macrocycles (especially cyclopeptides).

Several groups focused on the rational design and optimization of targeted, bioavailable *(bio)synthetic macrocycles* using modular and upscalable routes. Their *chemical synthesis via* macrocyclization was perceived as a difficult, substrate-dependent process.[209] Any macrocyclization competes against inter-molecular dimerization (Fig. 4.27).[210] Dilution should favor macrocyclization, but the process was often inefficient (especially with smaller, ≤13-membered macrocycles).[211]

Synthetic macrocyclic candidates required an upscalable synthesis. Examples were reported in literature,[212] and more should follow when other macrocycles will enter late preclinical evaluation. A macrocyclization efficiency index (EMAC[213]) evaluated the efficiency of ≈900 macrocyclization reactions. Reaction concentration (C) and yields (Y) contributed to EMAC, according to the equation

$$EMAC = \log_{10}\left[Y^3 \times C\right].$$

An EMACs ≥ 5 corresponded to macrocyclizations that in diluted conditions provided high yields in macrocycles. Macrocyclizations with EMACs ≥ 7 could be completed at higher concentrations (≈10 mM), while EMACs ≤ 3 yielded low yields (≤30%) even at high dilutions.[213]

A review[212] covered more than 10 macrocyclization strategies/reactions. Macrolactamization (cyclic peptides) and lactonization (macrolides), Sn2 and SnAr substitutions, click chemistry, ring-closing metathesis, Pd-catalyzed couplings, Wittig and Mitsunobu reactions, retro Diels Alder condensations, free radical chemistry, and multicomponent (MCR) reactions[212] led to synthetic

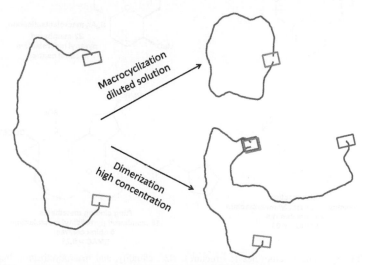

FIG. 4.27 Synthetic macrocycles: macrocyclization vs. dimerization.

macrocycles varying in ring size, compound classes, and achievable CD. Some are shown in Figs. 4.28 and 4.29 (boxed sites for macrocyclization).

214 synthetic macrocycles (including enantio- and diastereomeric pairs) with low RO5 compliance (71% with one, 34% with two noncompliant parameters) were studied.[214] Differences in passive cell permeability (Caco-2 adsorption,[215] apical-to-basolateral/AP and basolateral-to-apical/BA permeability) and in transporter-mediated efflux[216] (efflux ratio) were explained by positive (cLogP/-1 to 8 in the set) and negative factors (HBAs/5 to 15, HDBs/0 to 5 and PSA/60 to 250Å in the set). Substructural features impacted permeability

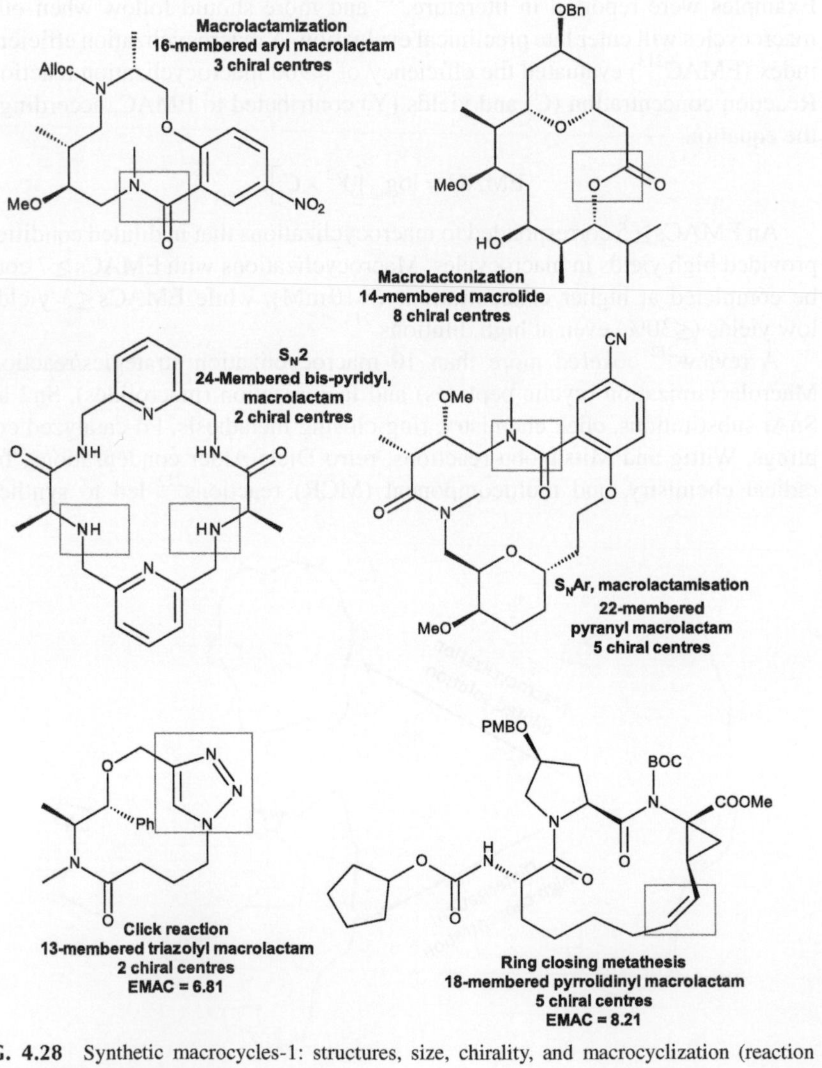

FIG. 4.28 Synthetic macrocycles-1: structures, size, chirality, and macrocyclization (reaction type, connections).

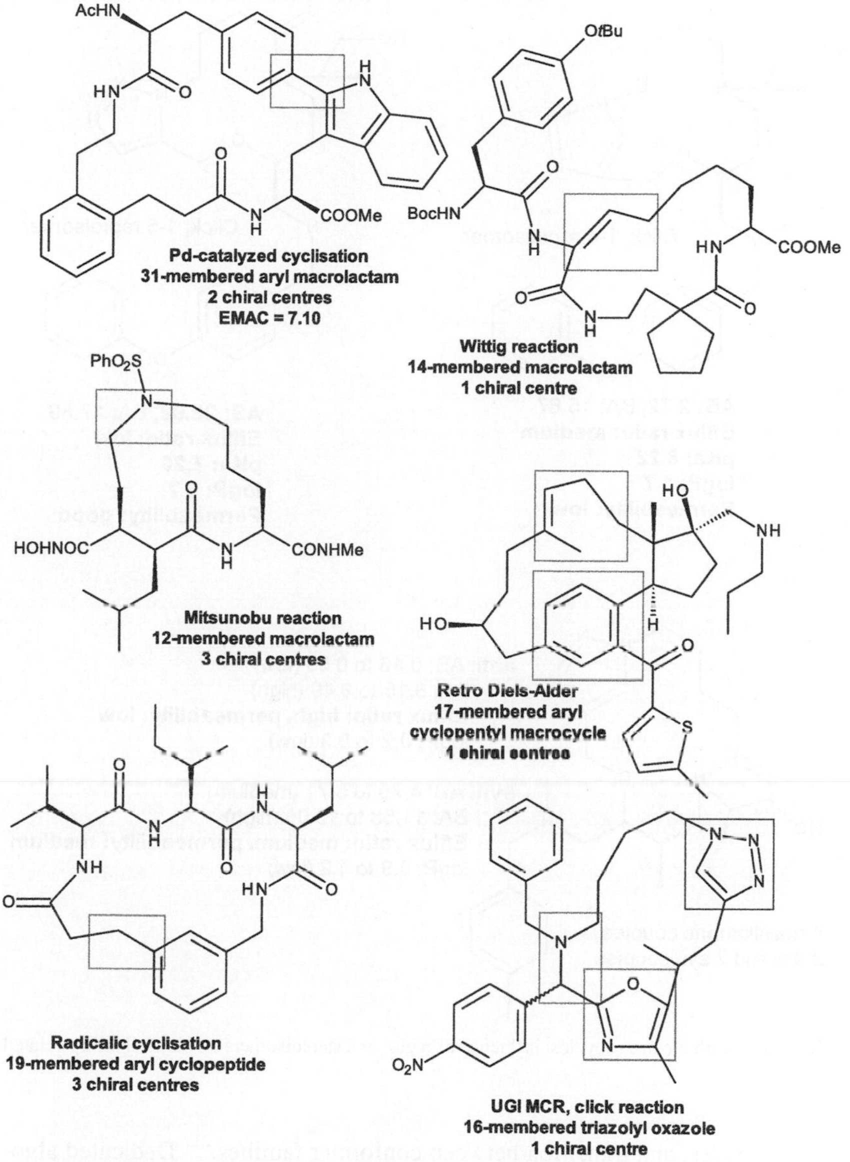

FIG. 4.29 Synthetic macrocycles-2: structures, size, chirality, and macrocyclization (reaction type, connections).

(increased by phenyl rings, pyridines, isoxazoles, and tertiary amines; decreased by ureas, carbonyls, sulfonamides, and secondary amines). Regio- and stereoisomers showed property-dependent permeability changes,[214] shown in Fig. 4.30.

Synthetic macrocycles should be in HTS collections. As to *virtual collections*, old computational methods could not sample the conformational space

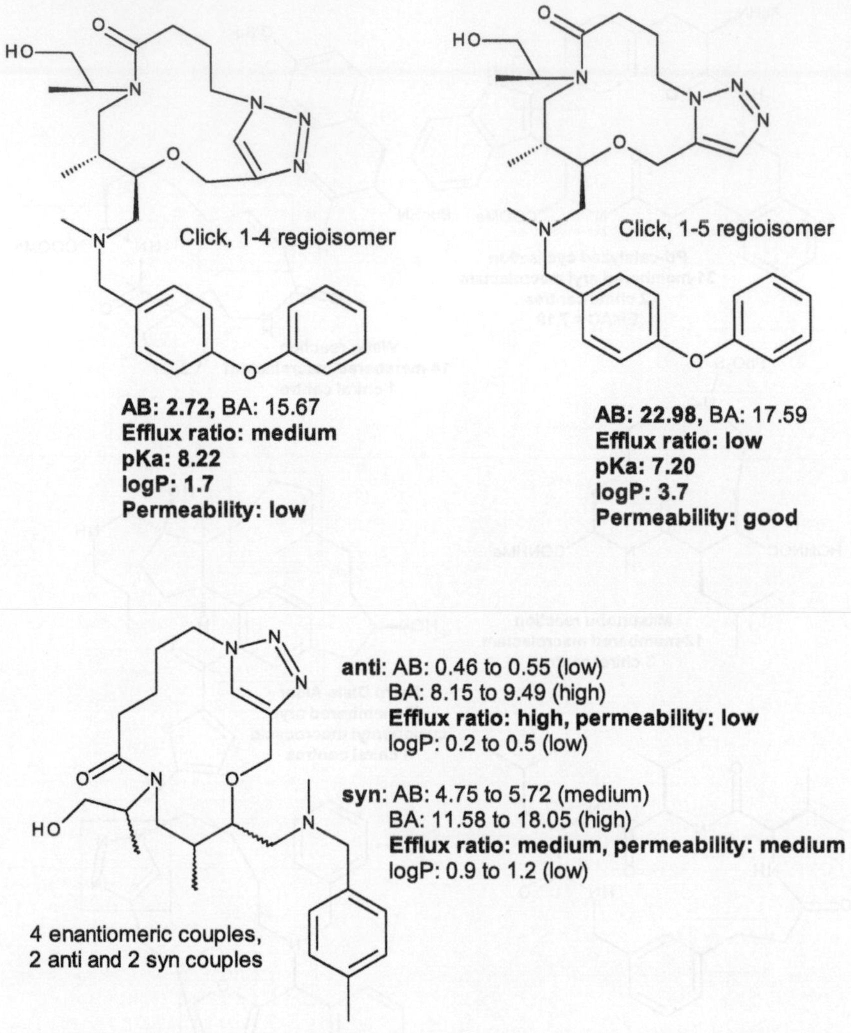

Click, 1-4 regioisomer

AB: 2.72, BA: 15.67
Efflux ratio: medium
pKa: 8.22
logP: 1.7
Permeability: low

Click, 1-5 regioisomer

AB: 22.98, BA: 17.59
Efflux ratio: low
pKa: 7.20
logP: 3.7
Permeability: good

anti: AB: 0.46 to 0.55 (low)
BA: 8.15 to 9.49 (high)
Efflux ratio: high, permeability: low
logP: 0.2 to 0.5 (low)

syn: AB: 4.75 to 5.72 (medium)
BA: 11.58 to 18.05 (high)
Efflux ratio: medium, permeability: medium
logP: 0.9 to 1.2 (low)

4 enantiomeric couples,
2 anti and 2 syn couples

FIG. 4.30 Synthetic macrocycles: influence of regio- and stereoisomers on bioavailability-related parameters.

of macrocycles, and transition between conformer families.[217] Dedicated algorithms and biophysics (i.e., NMR and X-ray) enabled reliable virtual models of macrocyclic families for vHTS, and to guide the optimization of macrocyclic hits.[218]

Tangible collections should contain *synthetic macrocycle libraries*. Cyclic peptide-based (i.e., *top*, Fig. 4.31) and macrolide-based libraries (i.e., *middle*) were made through macrocyclizations in solution or on SP.

Smaller libraries based on the build-couple-pair (B-C-P) strategy[219] yielded decorated nonpeptidic macrocycles (*bottom*, Fig. 4.31). Synthetic macrocycle libraries[212,219–221] were reviewed recently.

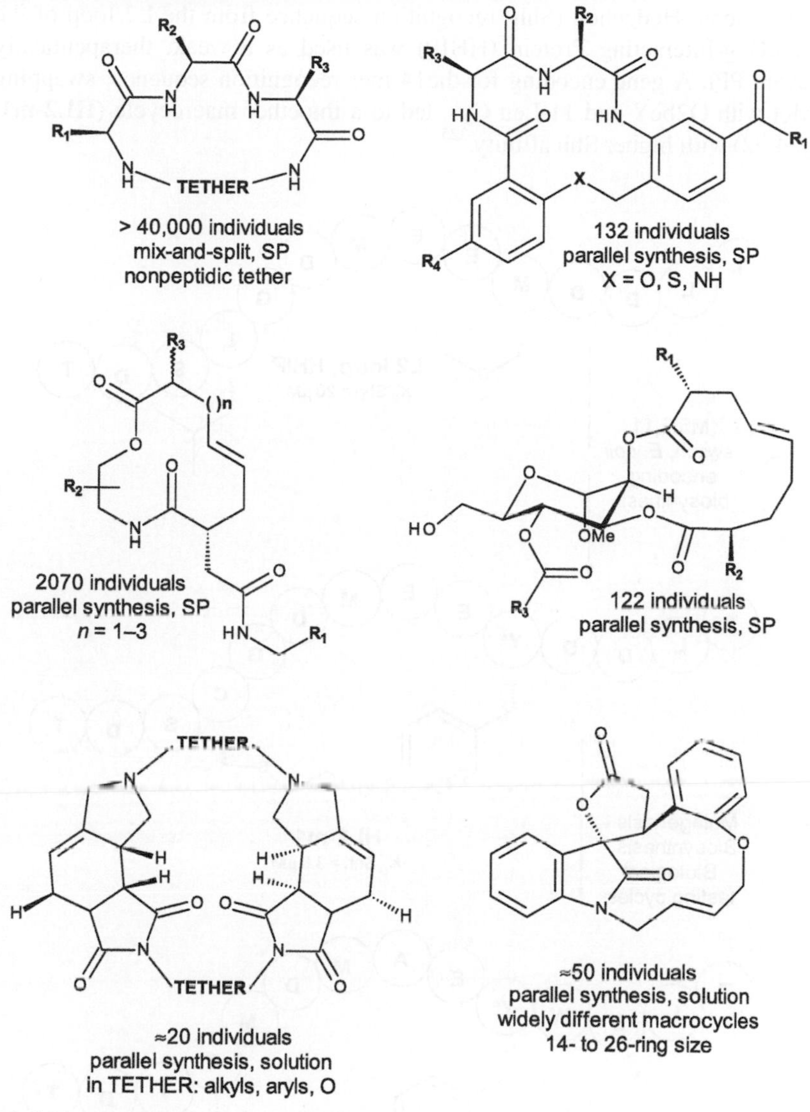

> 40,000 individuals
mix-and-split, SP
nonpeptidic tether

132 individuals
parallel synthesis, SP
X = O, S, NH

2070 individuals
parallel synthesis, SP
n = 1–3

122 individuals
parallel synthesis, SP

≈20 individuals
parallel synthesis, solution
in TETHER: alkyls, aryls, O

≈50 individuals
parallel synthesis, solution
widely different macrocycles
14- to 26-ring size

FIG. 4.31 Macrocyclic libraries: peptide-based (*top*), macrolide-based (*middle*), and miscellaneous libraries (*bottom*).

Biosynthetic macrocycles were obtained by methods at the interface between chemistry and biology.[209,222]

The incorporation of *O-2-bromoethyl tyrosine* (O2beY) in the genetic code of *E. coli* strains *via* an aminoacyl-tRNA synthetase (AARS)/tRNA pair led to the biosynthesis of O2beY-containing peptidomimetics.[223] Spontaneous intramolecular cyclization took place between O2beY and a Cys residue, with the help of a suitably placed Asp residue. Lanthionide-like biosynthetic macrocycles were obtained.[224]

The Sonic Hedgehog (Shh) recognition sequence from the L2 loop of the HedgeHog-Interacting Protein (HHIP) was used as a weak, therapeutically relevant PPI. A gene encoding for the 14-mer recognition sequence, swapping 5-Met with O2beY and 11-Leu Cys, led to a thioether macrocycle (**HL2-m1**, Fig. 4.32) with higher Shh affinity.[225]

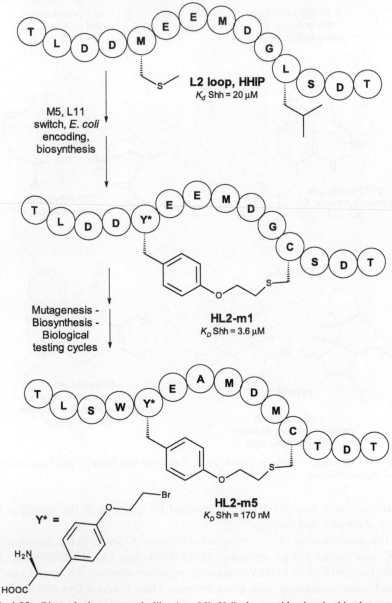

FIG. 4.32 Biosynthetic macrocycle libraries: O2beY-display peptidomimetic thioether macrocycle libraries and leads.

The O2beY method allowed little structural diversification with a low throughput. A focused array ($\approx$500 individuals) was made in a few biosynthesis-testing iterations, studying the influence of mutations on **H2L-m1**. Individuals were expressed with an affinity tag, purified and tested on immobilized Shh. The most potent library individual, **HL2-m5** (Fig. 4.32), showed a >100-fold affinity improvement compared with the linear recognition sequence. Large-scale synthesis of **H2L-m5** was carried out by SP synthesis. The O2beY method was suitable for the identification of biologically active, stable, constrained cyclic peptidomimetics, and for their focused structural optimization.

Modification of the *E. coli* genetic code *via* flexizime (flexible tRNA acylation ribozymes) technology[226] in a cell-free in vitro translation (FIT) system[227] was coupled to mRNA display,[228] providing the HT display method *RaPID* (*RAndom nonstandard Peptides Integrated Discovery*)[229] (Fig. 4.33).

The modified translational codon code (*top*, Fig. 4.33) entailed the replacement of five aminoacids with an alkylating aminoacid (2-ChloroacetylTyr, ClAcY), to ensure thioether cyclization with Cys residues, and with four N-methylated aminoacids, to increase cell permeability. Transcription of a cDNA library (step 1) and conjugation with puromycin (step 2) yielded a transcribed, puromycin-linked mRNA library. Library members were translated by the FIT system into puromycin-connected mRNA-peptidomimetic hybrids (step 3). Their peptidomimetic portion underwent spontaneous cyclization to thioether macrocycles (step 4). The thioether macrocycle library was tested as pooled

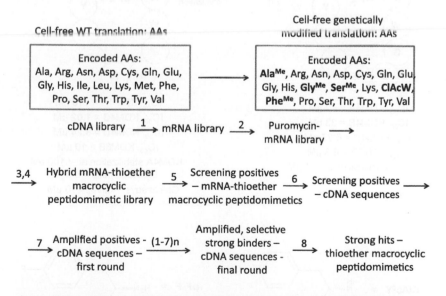

1: transcription; 2: conjugation with puromycin; 3: translation by FIT; 4: spontaneous cyclization; 5: affinity test, pooled mixtures; 6: reverse transcription, screening positives; 7: PCR amplification; 8: DNA decoding.

FIG. 4.33 Biosynthetic macrocycle libraries: RAPID-display peptidomimetic thioether macrocycle libraries, biosynthesis, and screening.

mixtures against immobilized molecular targets (step 5). Reverse transcription of positives into their cDNA counterparts (step 6) and PCR amplification (step 7) were followed by other transcription-translation-macrocyclization-selection rounds (steps 1 to 7, n times). All selected cDNA sequences were decoded (step 8, Fig. 4.33) to identify potent thioether peptidomimetic macrocycles.

RaPID was suited for HT identification of macrocyclic thioether peptidomimetics. Two $> 10^{12}$ mDNA libraries (L-/2ClAc-^{L}Y or D-/2ClAc-^{D}Y alkylating aminoacid) were screened against the catalytic domain of the KDM4 lysine demethylase subfamily.[230] Out of 44 hits, the thioether macrocycle **CP2** (Fig. 4.34) showed nM potency against KDM4A-C, high intra- ($>$100-fold potency of KDM4A-C vs. KDM4D-E) and interfamily selectivity ($>$ 100-fold vs. other lysine demethylases).

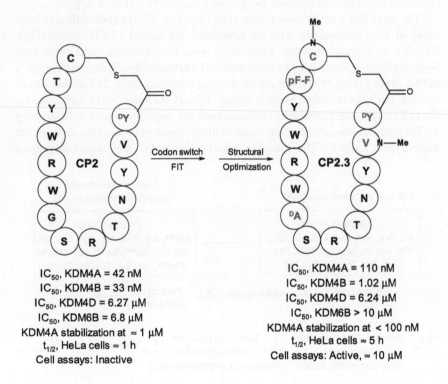

IC$_{50}$, KDM4A = 42 nM
IC$_{50}$, KDM4B = 33 nM
IC$_{50}$, KDM4D = 6.27 μM
IC$_{50}$, KDM6B = 6.8 μM
KDM4A stabilization at $\approx$ 1 μM
t$_{1/2}$, HeLa cells $\approx$ 1 h
Cell assays: Inactive

Codon switch FIT

Structural Optimization

IC$_{50}$, KDM4A = 110 nM
IC$_{50}$, KDM4B = 1.02 μM
IC$_{50}$, KDM4D = 6.24 μM
IC$_{50}$, KDM6B $>$ 10 μM
KDM4A stabilization at $<$ 100 nM
t$_{1/2}$, HeLa cells $\approx$ 5 h
Cell assays: Active, $\approx$ 10 μM

ClAc^{D}Y =

pF-F = H$_2$N

FIG. 4.34 RAPID-display peptidomimetic thioether macrocycle libraries: cell-active inhibitors of KDM4A-dependent lysine demethylation.

Cocrystallization of **CP2** with KDM4A explained its potency and selectivity. **CP2**-based probes clarified the role of KMN4A-dependent lysine demethylation. An array of **CP2** analogues (unnatural aminoacids—stabilization to proteolysis; N-methylation → cell permeability, stabilization to proteolysis; poly-Arg chains → cellular uptake) contained leads endowed with cellular activity (i.e., **CP2.3**, Fig. 4.34[230]).

DNA-encoded chemistry and affinity screening[231,232] was used to prepare and screen macrocycle libraries.[219,222,233] The optimization of hits from DNA-encoded libraries led to the development and clinical testing of a candidate.[234] DNA-encoded chemical libraries (DECLs) were introduced in the early '90s.[235] Their synthesis used mix-and-split protocols[14] in solution, where oligonucleotide (ON) chains encoded each synthetic step (Fig. 4.35).

ON sequences were functionalized with a linker (L) ending with a reactive, DNA-compatible moiety. The DNA-L construct was split into n vessels, and coupled through L with the first set of reagents (**A1-n**, step 1, Fig. 4.35). Each vessel was treated with one of the n **A1-n** reagents, and each DNA-L-A_x intermediate was reacted with the corresponding oligo sequence/code from an oligo codes **A1-n** (enzymatic[236] or chemical ligation,[237] step 2). The n-membered intermediate library was pooled in a vessel (step 3). An identical split-couple-encode sequence (steps 4, 5) was repeated in m vessels for a second set of reagents (**B1-m**, Fig. 4.35), producing after pooling (step 6) a (n x m), two building blocks (2-BB) DECL. Published libraries were made by up to 8 couplings (8-BBs),[238] and by up to billions of individuals.[239]

Alternative synthetic protocols for smaller DECLs exploited DNA sequences to guide the reactions among BBs. In *DNA-templated chemistry* (*DTC*[240]), libraries were built from two complementary codon-anticodon single-stranded (SS) DNA-BBs' libraries. After hybridization, BB pairs were brought close increasing their local concentration and reactivity.[241] The anticodon DNA library was bound to its BBs by a cleavable linker, so that its reaction with codon-bound BBs led after linker cleavage to an encoded, codon DNA-bound DECL. Up to 160,000-membered DTC libraries were reported.[242]

SS DNA complementarity was exploited through multiple SS DNA interactions, to form 3- or 4-way DNA junctions. Three or four BBs were brought close, and the corresponding 3- or 4-BB DECLs were synthesized. The volume of the DNA junction was estimated as $\approx 10^{-24}$ L, i.e., a yoctoliter—thus the *yoctoReactor* (y^{R243}) name. Stepwise assembly of the DNA junction[244] and concerted assembly *via* MCRs were reported.[245] Up to 12.6 million-membered y^R libraries were published.[244]

Encoded self-assembling libraries (*ESACs*[246]) were built by hybridization of complementary SS DNA libraries, independently synthesized and functionalized with BBs at opposite poles. The BBs linked to each hybridization pair were coupled through a linker that could contribute to the potency of active library individuals.[247] Up to 110,000-membered ESAC libraries were reported.[248]

The strategy for DECL screening is shown in Fig. 4.36.

DECLs were tested as single pools, measuring the affinity of each library member for the target.[249] The process resembles display methods, such as phage

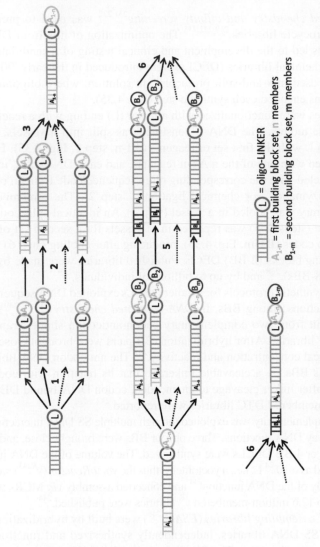

1: split in n reactors, coupling reaction, building blocks A_1 to A_n; 2: ligation, oligo codes A_1 to A_n; 3: pooling;
4: split in m reactors, coupling reaction, building blocks B_1 to B_m; 5: ligation, oligo codes B_1 to B_m; 6: pooling.

L = oligo-LINKER
A_{1-n} = first building block set, n members
B_{1-m} = second building block set, m members

FIG. 4.35 DNA-encoded libraries (DECLs): mix-and-split synthesis in solution.

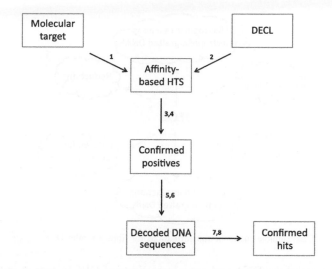

1: target immobilization (preferred option); 2: DECL incubation with target; 3: 1st selection round; 4: other selection round(s) (optional); 5: PCR amplification; 6: HT DNA sequencing; 7: synthesis of active hits; 8: hit profiling.

FIG. 4.36   DECLs: screening strategy.

display. The target was often immobilized on solid surfaces (magnetic beads,[250] resins,[251] agarose[252]—step 1, Fig. 4.36). Solubilized targets were used when immobilization affected target functionality,[231,253] A first selection round incubated the DECLs with the target (step 2), washed unbound library members, and eluted target-binding positives (step 3). Affinity-based screening simplified the assay protocol and reduced the target needed.[250] Multiple selection rounds (step 4) were needed when many binders were found, to focus onto strong, specific binders. PCR amplification (step 5) and DNA sequencing (step 6) of the DNA codes hits provided the structure of the target-binding positives. They were synthesized as discrete, noncoded compounds (step 7) and were carefully profiled in vitro (step 8, Fig. 4.36). Positives caused by DNA chain-dependent target affinity, or by binding to epitopes, which did not affect target functionality, were discarded.

DNA is a delicate molecule (limited acid stability and organic solvent solubility), and its use in DNA-encoded chemistry limited the *chemical transformations* compatible with DECLs[254] (Fig. 4.37).

Amide formation on amine-grafted (carboxylates as BBs) or electrophile-grafted DNA (amines as BBs) was the most frequent reaction, followed by nucleophilic substitutions and reductive aminations. Deprotection protocols for amine- (i.e., Boc, Fmoc, and Alloc) and acid-protecting groups (i.e., Me and t-Bu esters) enabled the use of multifunctional BBs. Reductions (i.e., nitro and azide groups), C-C couplings (i.e., Suzuki-Miyaura and Wittig reactions) and cycloadditions (i.e., Diels-Alder and Huysgen/click cycloadditions) were also used in DECL synthesis. A review listed >40 assessed chemical transformations for DNA-encoded chemistry[255]; another covered DNA chemistry in organic media.[256]

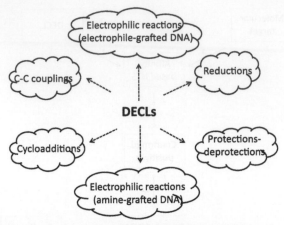

FIG. 4.37 DECLs: suitable chemical transformations for library synthesis.

The *quality* of the DNA and organic portions of DECLs was monitored,[257] without complete postsynthesis analytical characterization of large DECL libraries. The efficiency of each reaction step was assessed before DECL synthesis on individual model reactions (*reaction rehearsal*[255]); BBs' reactivity was evaluated on model reagents (monomer rehearsal[258]). Excess /unreacted reagents were removed by EtOH precipitation or ion-exchange chromatography of intermediate DNA-organic constructs.[259] Unreacted intermediates were capped with DNA-compatible reagents after DECL synthesis.[260] QC on reaction mixtures was carried out by gel electrophoresis (checking DNA codes[251]), HPLC-MS (checking DNA-BB constructs used during DECL synthesis[261]), and ion pair HPLC (checking pools of DNA-multiBB constructs[262]).

The quality of DECL libraries was correlated with their screening performance. Computational simulations showed that poorly represented individuals (lower chemical yields) often behaved as false negatives in a screening campaign,[263] mostly when large DECLs were used.[264]

DECLs could be built with linear reaction sequences, where typically one (2-BB DECLs) to three bifunctional BBs (4-BB DECLs) were sequentially coupled and capped with a monofunctional BB (*top* lane, Fig. 4.38). Scaffold-like, tri- (*left*, *bottom* lane) or tetrafunctional BBs (*right*, *bottom* lane, Fig. 4.38) could be sequentially capped with two or three monofunctional BBs in a 3- or 4-BB DECL, respectively.

Commercially available monofunctional reagents suitable for DECL synthesis are large, diverse sets. Sets commercially available bifunctional reagents (needing a DNA-compliant protecting group) are smaller, and even less trifunctional reagents can be purchased. With amide formation as an example, >20 K monofunctional carboxylic acids and amines could be purchased at affordable prices, while only a few hundreds of protected bifunctional aminoacids (and even less trifunctional aminoacids) could be purchased at low cost.[265]

The diversity and the size of a DECL are mostly dependent on mono- and multi- (>2) functional BBs. A multifunctional BB/scaffold capped with

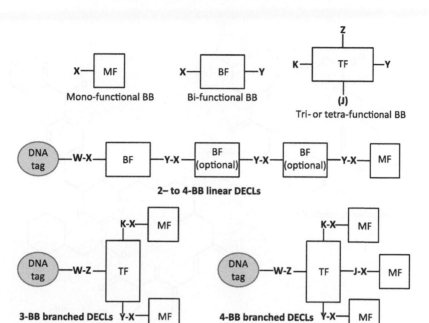

FIG. 4.38 DECLs: scaffolds, BBs, and build-up strategies.

monofunctional BBs provides more spatial orientations/tridimensional geometries (to explore the CD space suitable to interact with binding pockets of molecular targets) than a string of bifunctional BBs in a linear DECL.[254] Thus, synthetic efforts aimed to multifunctional scaffolds onto which to build branched DECLs should be prioritized.[232]

DECLs varied in terms of drug-likeness.[231,254] Assuming an upper MW = 200 limit for BBs, a 3-BB DECL is close to exceeding the upper MW limit for RO5-compliant/drug-like small molecules. A 4-BB DECL would contain a majority of non-RO5-compliant library members, whose positives after a selection process would be tough to optimize. Many commercially available, low MW monofunctional BBs allowed the access to diverse and large RO5-compliant 3-BB DECL (i.e., a 2-M branched 3-BB DECL made by reacting 50 trifunctional BBs and two 200-membered, low MW sets of monofunctional BBs).

2-BB DECLs yielded hits against difficult PPI targets,[266,267] but larger 3- and 4-BB DECLs were more successful against noncanonical targets. *Macrocyclic DECLs*[233] provided hits against the cytokine IL17 (*left*[268]) and insulin-degrading enzyme IDE (*right*[269]) shown in Fig. 4.39.

The PPI between Inhibitor of Apoptosis proteins (IAPs) and their endogenous ligand Smac[270] is validated against cancer and autoimmune diseases.[271] IAPs bind caspases through their Baculovirus IAP Repeat (BIR) domains BIR2 and BIR3, and regulate caspase-dependent apoptotic pathways.[272] Cells overexpressing IAPs, and X-linked IAP (XIAP) in particular, become more resistant to apoptosis.[273]

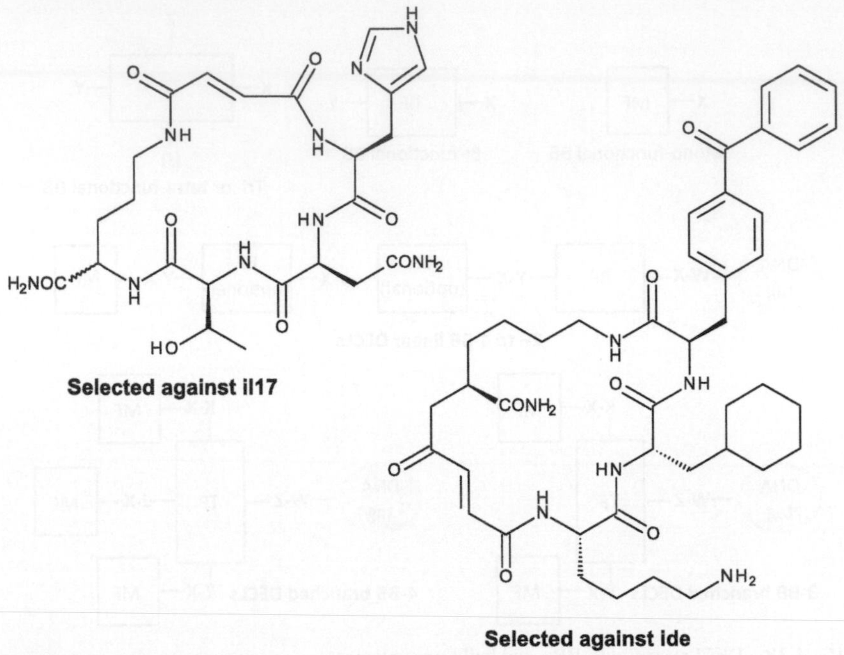

Selected against il17

Selected against ide

FIG. 4.39 Macrocyclic DECLs: structures of hits against IL17 (*left*) and IDE (*right*).

Monomeric and dimeric Smac mimetics binding with BIR3, or with both BIR2 and BIR3 of IAPs, respectively, entered clinical development.[274,275] Monomers prevented BIR3-driven binding with initiator caspase 9 by mimicking the Smac tetrapeptide AVPI sequence (*top*, Fig. 4.40). Dimeric Smac mimics blocked also the BIR2-mediated interaction between IAPs and effector caspases.

A 160,000-membered, 5-BB macrocyclic DECL (*bottom*, Fig. 4.40) targeted the Smac-IAP PPI by mimicking N-terminal AVPI.[244] Twenty aminoacids were selected for the P_1 (Ala), P_3 (Pro), and P_4 (Ile) position; four alkyne-containing aminoacids were chosen for P_2 (Val); and five azide-containing aminoacids were selected for P_5 (Ala in Smac). The linear precursor library was built using DTC protocols,[276] and macrocyclization was achieved by Huysgen cycloaddition/click chemistry (Fig. 4.40).

The P_2-P_5 macrocyclic library, partitioned into four 40,000-membered pools varying for their P_2 position, was tested as shown in Fig. 4.41, *top*.

Positives were selected after incubation with either His-tagged BIR3 immobilized on nickel resin (step 1a), or biotinylated BIR2 immobilized on streptavidin beads (step 1b). After washout of nonbound macrocycles (step 2), denaturation and elution (step 3) led to the isolation of DNA-macrocycle positives. PCR amplification (step 4) and sequencing (step 5) led to the DTC code for each positive. The BBs' preferences for most of the library positions (P_1 to P_4) and the structure

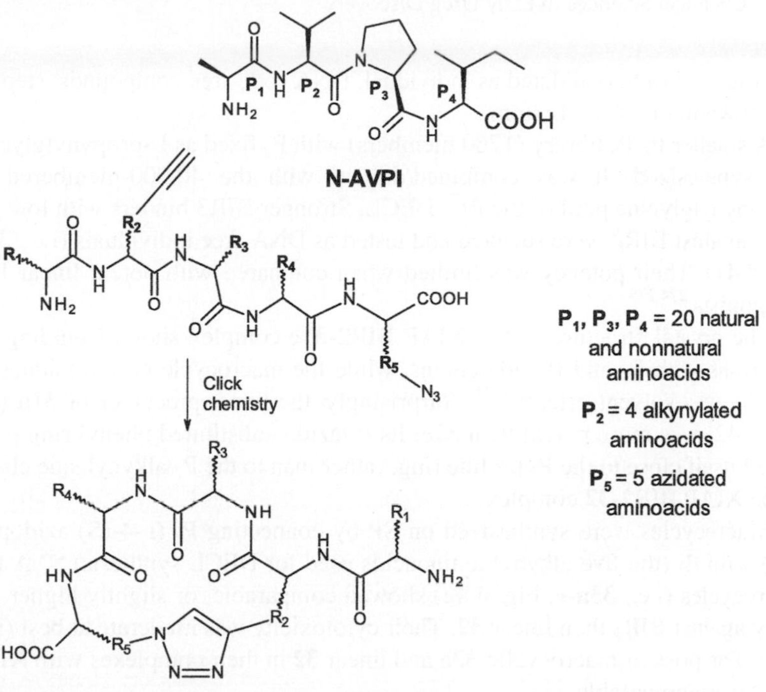
N-AVPI
P₁, P₃, P₄ = 20 natural and nonnatural aminoacids
P₂ = 4 alkynylated aminoacids
P₅ = 5 azidated aminoacids
Click chemistry

FIG. 4.40 IAP antagonists: structures of the AVPI tetrapeptide model (*top*) and of the 160,000-membered macrocyclic DECL (*bottom*).

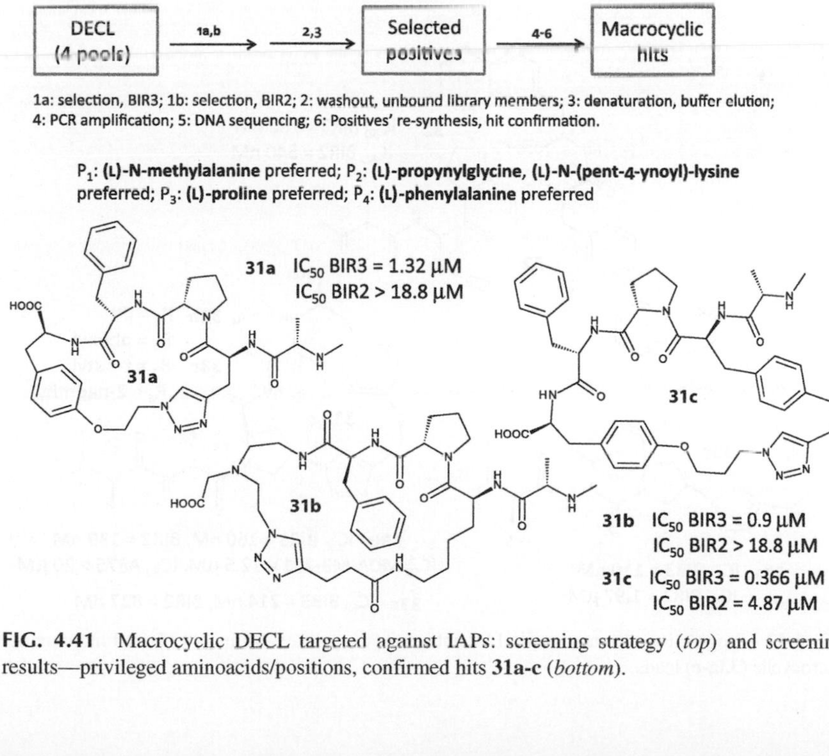
DECL (4 pools)
1a,b
2,3
Selected positives
4-6
Macrocyclic hits
1a: selection, BIR3; 1b: selection, BIR2; 2: washout, unbound library members; 3: denaturation, buffer elution; 4: PCR amplification; 5: DNA sequencing; 6: Positives' re-synthesis, hit confirmation.
P1: (L)-N-methylalanine preferred; P2: (L)-propynylglycine, (L)-N-(pent-4-ynoyl)-lysine preferred; P3: (L)-proline preferred; P4: (L)-phenylalanine preferred
31a IC50 BIR3 = 1.32 μM
IC50 BIR2 > 18.8 μM
31a
31b
31c
31b IC50 BIR3 = 0.9 μM
IC50 BIR2 > 18.8 μM
31c IC50 BIR3 = 0.366 μM
IC50 BIR2 = 4.87 μM

FIG. 4.41 Macrocyclic DECL targeted against IAPs: screening strategy (*top*) and screening results—privileged aminoacids/positions, confirmed hits **31a-c** (*bottom*).

of positives **31a,b**, (validated as individual, DNA code-free compounds, step 6), are shown in Fig. 4.41, bottom.

A smaller P_2-P_5 library (1760 members) with P_2 fixed as L-propynylglycine was synthesized. It was combined/diluted with the 40,000-membered P_2 L-propynylglycine pool of the first DECL. Stronger BIR3 binders with low potency against BIR2 were selected and tested as DNA-free individuals (i.e., **31c**, Fig. 4.41). Their potency was limited when compared with potent linear IAP antagonists.[274,275]

The crystal structure of the XIAP BIR2-**31c** complex showed binding interactions with P_1 and P_4 side chains, while the macrocycle core (residues P_2 and P_5) was solvent oriented.[244] Surprisingly, the linear precursor of **31a** (**32**, Fig. 4.42) was more potent than **31c**. Its P_5/azido-substituted phenyl ring positioned itself close to the P_3 proline ring, rather than to the P_2/alkynyl side chain, in the XIAP BIR2-**32** complex.

Macrocycles were synthesized on SP by connecting P_3 (L-4-(S)-azidoproline) with P_5 (the five alkynyl aminoacids used for DECL synthesis).[244] A few macrocycles (i.e., **33a-c**, Fig. 4.42) showed comparable, or slightly higher potency against BIRs than linear **32**. Their cytotoxicity was moderate at best (i.e., **33b**). The pose of macrocyclic **33a** and linear **32** in their complexes with XIAP BIR2 was comparable.

Dimeric Smac mimetics[277] bind to the XIAP BIR2-linker-BIR3 construct with higher affinity than monomers. Head-to-tail dimerization of P_3-P_5 macrocycles led to potent dimeric macrocycles (**34a,b**, Fig. 4.43).

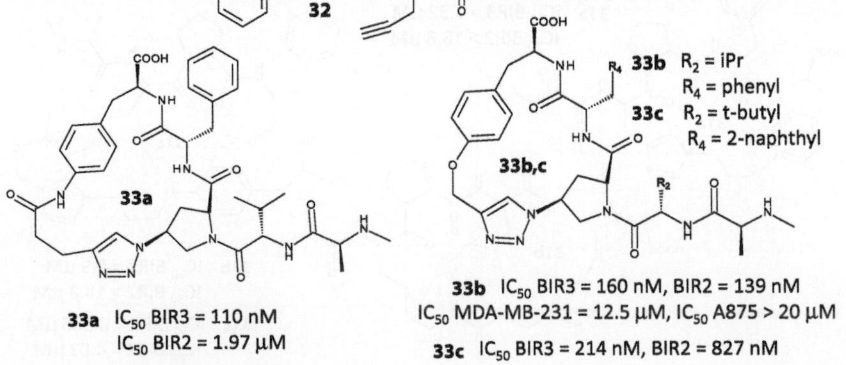

32 IC_{50} BIR3 = 180 nM
IC_{50} BIR2 = 540 nM

33b R_2 = iPr
R_4 = phenyl
33c R_2 = t-butyl
R_4 = 2-naphthyl

33a IC_{50} BIR3 = 110 nM
IC_{50} BIR2 = 1.97 μM

33b IC_{50} BIR3 = 160 nM, BIR2 = 139 nM
IC_{50} MDA-MB-231 = 12.5 μM, IC_{50} A875 > 20 μM
33c IC_{50} BIR3 = 214 nM, BIR2 = 827 nM

FIG. 4.42 Progression of macrocyclic IAP antagonists: structures of linear (**32**) and monomeric macrocyclic (**33a-c**) leads.

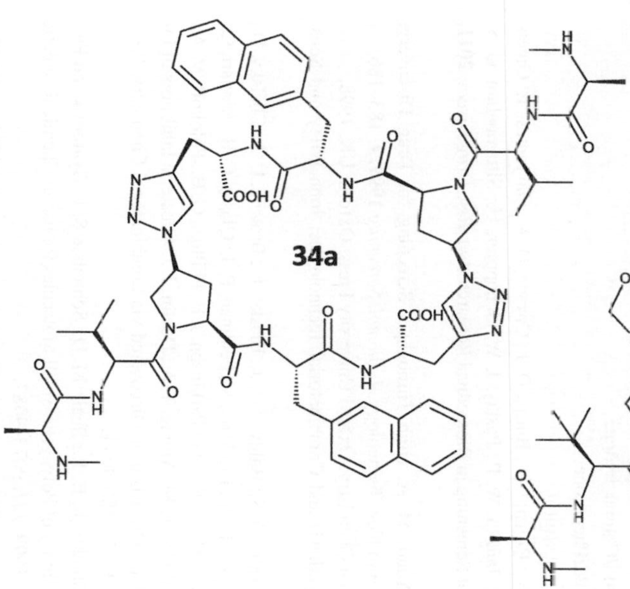

34b R_1 = OH, R_2 = 1-4-bound triazole

34c R_1 = NHSO$_2$cyclopropyl, R_2 = (CH$_2$)$_3$

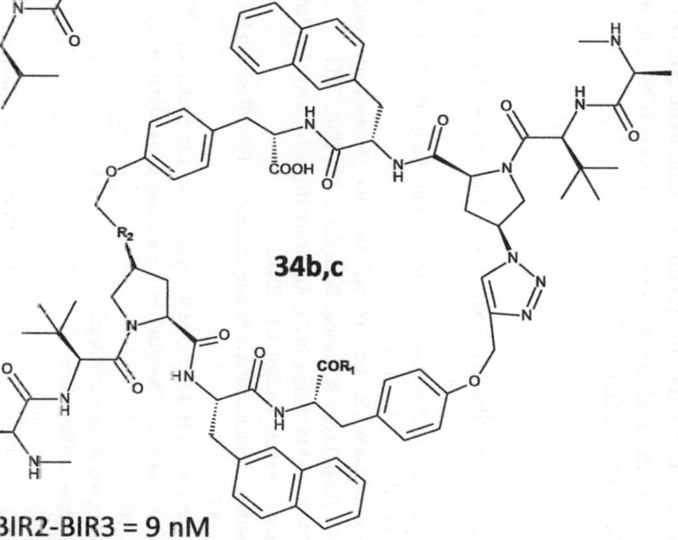

34a IC$_{50}$ BIR3 = 65 nM, BIR2 = 300 nM, BIR2-BIR3 = 9 nM
IC$_{50}$ MDA-MB-231 = 2.01 µM, IC$_{50}$ A875 = 6.14 µM

34b IC$_{50}$ BIR3 = 36 nM, BIR2 = 97 nM, BIR2-BIR3 = 22 nM
IC$_{50}$ MDA-MB-231 = 26 nM, IC$_{50}$ A875 = 77 nM

34c IC$_{50}$ BIR3 = 3.5 nM, BIR2-BIR3 = 0.8 nM, IC$_{50}$ A875 = 19 nM

FIG. 4.43 Progression of macrocyclic IAP antagonists: structures of dimeric macrocyclic leads (**34a,b**) and of preclinical candidate **34c**.

The lead **34b** showed good cytotoxicity, a reasonable PK/PD profile by i.v. administration, and moderate efficacy (50 mg/kg) in a murine xenograft model (MDA-MB-231 breast cancer cells).[244]

Further lead optimization[278] led to the preclinical candidate **34c** (Fig. 4.43), endowed with excellent in vitro and in vivo activity (2 mg/kg, 67% tumor growth inhibition/TGI; 5 mg/kg, 100% TGI).

REFERENCES

1. Lietava, J. Medicinal Plants in a Middle Paleolithic Grave Shanidar IV? *J. Ethnopharmacol.* **1992**, *35*, 263–266.
2. Gorman, C. F. Hoabinhian: A Pebble-Tool Complex with Early Plant Associations in Southeast Asia. *Science* **1969**, *163*, 671–673.
3. Benzie, I. F. F., Wachtel-Galor, S., Eds. *Herbal Medicine: Biomolecular and Clinical Aspects*; 2nd ed.; CRC Press/Taylor & Francis: Boca Raton, FL, USA, 2011.
4. Rider, C. Religion, Magic and Medicine. In *The Routledge History of Medicine*; Jackson, M., Ed.; Routledge/Taylor & Francis: Oxford, UK, 2016.
5. Vickery, B. W. Scientific Communication in History; Scarecrow Press/Rowman & Littlefield: Lanham, MD, USA, 2000.
6. Sneader, W. Drug Discovery: A History; John Wiley and Sons: Chichester, UK, 2005.
7. Greenberg, A. From Alchemy to Chemistry in Picture and Story; John Wiley and Sons: Hoboken, NJ, USA, 2007.
8. https://en.wikipedia.org/wiki/Chloral_hydrate.
9. https://en.wikipedia.org/wiki/Phenazone.
10. https://en.wikipedia.org/wiki/Aspirin.
11. Macarron, R.; Banks, M. N.; Bojanic, D.; Burns, D. J.; Cirovic, D. A.; Garyantes, T.; Green, D. V. S.; Hertzberg, R. P.; Janzen, W. P.; Paslay, J. W.; Schoppfer, U.; Sittampalam, G. S. Impact of High Throughput Screening in Biomedical Research. *Nat. Rev. Drug Discov.* **2011**, *10*, 188–195.
12. Fox, S.; Farr-Jones, S.; Yund, M. A. High Throughput Screening for Drug Discovery: Continually Transitioning into New Technology. *J. Biomol. Screening* **1999**, *4*, 183–186.
13. Terrett, N. K. Combinatorial Chemistry; Oxford University Press: Oxford, UK, 1998.
14. Seneci, P. Solid Phase Synthesis and Combinatorial Technologies; John Wiley and Sons: New York, US, 2000.
15. McMillan, K.; Adler, M.; Auld, D. S.; Baldwin, J. J.; Blasko, E.; Browne, L. J.; Chelsky, D.; Davey, D.; Dolle, R. E.; Eagen, K. A.; Erickson, S.; Feldman, R. I.; Glaser, C. B.; Mallari, C.; Morrissey, M. M.; Ohlmeyer, M. H.; Pan, G.; Parkinson, J. F.; Phillips, G. B.; Polokoff, M. A.; Sigal, N. H.; Vergona, R.; Whitlow, M.; Young, T. A.; Devlin, J. J. Allosteric Inhibitors of Inducible Nitric Oxide Synthase Dimerization Discovered Via Combinatorial Chemistry. *Proc. Natl. Acad. Sci. U S A* **2000**, *97*, 1506–1511.
16. Tan, D. S.; Foley, M. A.; Stockwell, B. R.; Shair, M. D.; Schreiber, S. L. Synthesis and Preliminary Evaluation of a Library of Polycyclic Small Molecules for use in Chemical Genetic Assays. *J. Am. Chem. Soc.* **1999**, *121*, 9073–9087.
17. Weller, H. N. Purification of Combinatorial Libraries. *Molecular Diversity* **1999**, *4*, 47–52.
18. Baldino, C. M.; Caserta, J.; Goetzinger, W.; Harris, M.; Hartsough, D.; Yohannes, D.; Yu, L.; Kyranos, J. N. Accelerating drug discovery through automated parallel chemistry. *G.I.T. Lab. Journal* **2006**, *10*, 42–45.

19. Powers, D. G.; Casebier, D. S.; Fokas, D.; Ryan, W. J.; Troth, J. R.; Coffen, D. L. Automated Parallel Synthesis of Chalcone-Based Screening Libraries. *Tetrahedron* **1998**, *54*, 4085–4096.

20. Merritt, A. T. Solution-Phase Combinatorial Chemistry. *Combi. Chem. High Throughput Chem.* **1998**, *1*, 57–72.

21. Tropsha, A.; Zheng, W. Rational Principles of Compound Selection for Combinatorial Library Design. *Combi. Chem. High Throughput Chem.* **2002**, *5*, 111–123.

22. Shuttleworth, S. J.; Connors, R. V.; Fu, J.; Liu, J.; Lizarzaburu, M. E.; Qiu, W.; Sharma, R.; Wanska, M.; Zhang, A. J. Design and Synthesis of Protein Superfamily-Targeted Chemical Libraries for Lead Identification and Optimization. *Curr. Med. Chem.* **2005**, *12*, 1239–1281.

23. Prien, O. Target-Family-Oriented Focused Libraries for Kinases—Conceptual Design Aspects and Commercial Availability. *ChemBioChem* **2005**, *6*, 500–505.

24. Akritopoulou-Zanze, I.; Hajduk, P. J. Kinase-Targeted Libraries: The Design and Synthesis of Novel, Potent, and Selective Kinase Inhibitors. *Drug Discov. Today* **2009**, *14*, 291–297.

25. Turk, B. Targeting Proteases: Successes, Failures and Future Prospects. *Nat. Rev. Drug Discov.* **2006**, *5*, 785–799.

26. Lowrie, J. F.; Delisle, R. K.; Hobbs, D. W.; Diller, D. J. The Different Strategies for Designing GPCR and Kinase Targeted Libraries. *Combi. Chem. HTS* **2004**, *7*, 495–510.

27. Jacoby, E. Designing Compound Libraries Targeting GPCRs. *Ernst Schering Found Symp. Proc.* **2007**, *2*, 93–103.

28. Mestres, J.; Martín-Couce, L.; Gregori-Puigjané, E.; Cases, M.; Boyer, S. Ligand-Based Approach to in Silico Pharmacology: Nuclear Receptor Profiling. *J. Chem. Inf. Model.* **2006**, *46*, 2725–2736.

29. Harris, C. J.; Hill, R. D.; Sheppard, D. W.; Slater, M. J.; Stouten, P. F. W. The Design and Application of Target-Focused Compound Libraries. *Combi. Chem. HTS* **2011**, *14*, 521–531.

30. Gregori-Puigjané, E.; Mestres, J. Coverage and Bias of Chemical Libraries. *Curr. Opin. Chem. Biol.* **2008**, *12*, 359–365.

31. Lipinski, C. A.; Lombardo, F.; Dominy, B. W.; Feeney, P. J. Experimental and Computational Approaches to Estimate Solubility and Permeability in Drug Discovery and Development Settings. *Adv. Drug Delivery Rev.* **1997**, *23*, 3–25.

32. Cohen, B. E.; Bangham, A. D. Diffusion of Small Non-Electrolytes Across Liposome Membranes. *Nature* **1972**, *236*, 173–174.

33. Matsson, P.; Kihlberg, J. How Big Is Too Big for Cell Permeability? *J. Med. Chem.* **2017**, *60*, 1662–1664.

34. Moriguchi, I.; Hirono, S.; Nakagome, I.; Hiram, H. Comparison of Log P Values for Drugs Calculated by Several Methods. *Chem. Pharm. Bull.* **1994**, *42*, 976–978.

35. Abraham, M. H. Hydrogen Bonding. 31. Construction of A Scale of Solute Effective or Summation Hydrogen-Bond Basicity. *J. Phys. Org. Chem.* **1993**, *6*, 660–684.

36. Raevsky, O. A.; Grigor'ev, V. Y.; Kireev, D. B.; Zefirov, N. S. Complete Thermodynamic Description of Hbonding in the Framework of Multiplicative Approach. *Quant. Struct. Act. Relat.* **1992**, *111*, 49–63.

37. Veber, D. F.; Johnson, S. R.; Cheng, H.-Y.; Smith, B. R.; Ward, K. W.; Kopple, K. D. Molecular Properties That Influence the Oral Bioavailability of Drug Candidates. *J. Med. Chem.* **2002**, *45*, 2615–2623.

38. Petit, J.; Meurice, N.; Kaiser, C.; Maggiora, G. Softening the Rule of Five—Where to Draw the Line? *Bioorg. Med. Chem.* **2012**, *20*, 5343–5351.

39. Yusof, I.; Segall, M. D. Considering the Impact Drug-Like Properties Have on the Chance of Success. *Drug Discov. Today* **2013**, *18*, 659–666.

40. Lipinski, C. A. Rule of Five In 2015 and Beyond: Target and Ligand Structural Limitations, Ligand Chemistry Structure and Drug Discovery Project Decisions. *Adv. Drug Delivery Rev.* **2016**, *101*, 34–41.

41. Surade, S.; Blundell, T. L. Structural Biology and Drug Discovery of Difficult Targets: The Limits of Ligandability. *Chem. Biol.* **2012**, *19*, 42–50.

42. Doak, B. C.; Zheng, J.; Dobritzsch, D.; Kihlberg, J. How Beyond Rule of 5 Drugs and Clinical Candidates Bind to Their Targets. *J. Med. Chem.* **2016**, *59*, 2312–2327.

43. Baell, J. B.; Holloway, G. A. New Substructure Filters for Removal of Pan Assay Interference Compounds (PAINS) From Screening Libraries and for Their Exclusion in Bioassays. *J. Med. Chem.* **2010**, *53*, 2719–2740.

44. Eglen, R. M.; Reisine, T.; Roby, P.; Rouleau, N.; Illy, C.; Bossé, R.; Bielefeld, M. The Use of AlphaScreen Technology in HTS: Current Status. *Curr. Chem. Genomics* **2008**, *1*, 2–10.

45. Thorne, N.; Auld, D. S.; Inglese, J. Apparent Activity in High-Throughput Screening: Origins of Compound-Dependent Assay Interference. *Curr. Opin. Chem. Biol.* **2010**, *14*, 315–324.

46. Baell, J.; Walters, M. A. Chemical Con Artists Foil Drug Discovery. *Nature* **2014**, *513*, 481–483.

47. Dahlin, J. L.; Walters, M. A. How to Triage Pains-Full Research. *Assay Drug Dev. Technol.* **2016**, *14*, 168–174.

48. Baell, J. B. Feeling Nature's PAINS: Natural Products, Natural Product Drugs, and Pan Assay Interference Compounds (PAINS). *J. Nat. Prod.* **2016**, *79*, 616–628.

49. Aldrich, C.; Bertozzi, C.; Georg, G. L.; Kiessling, L.; Lindsey, C.; Liotta, D.; Merz, K. M., Jr.; Schepartz, A.; Wang, S. The Ecstasy and Agony of Assay Interference Compounds. *ACS Med. Chem. Lett.* **2017**, *8*, 379–382.

50. Capuzzi, S. J.; Muratov, E. N.; Tropsha, A. Phantom PAINS: Problems with the Utility of Alerts for Pan-Assay INterference CompoundS. *J. Chem. Inf. Model.* **2017**, *57*, 417–427.

51. Jasial, S.; Hu, Y.; Bajorath, J. How Frequently Are Pan-Assay Interference Compounds Active? Large-Scale Analysis of Screening Data Reveals Diverse Activity Profiles, Low Global Hit Frequency, and Many Consistently Inactive Compounds. *J. Med. Chem.* **2017**, *60*, 3879–3886.

52. Gilberg, E.; Jasial, S.; Stumpfe, D.; Dimova, D.; Bajorath, J. Highly Promiscuous Small Molecules from Biological Screening Assays Include Many Pan-Assay INterference compoundS but Also Candidates for Polypharmacology. *J. Med. Chem.* **2016**, *59*, 10285–10290.

53. Erlanson, D. Introduction to Fragment-Based Drug Discovery. *Top. Curr. Chem.* **2012**, *317*, 1–32.

54. Congreve, M.; Carr, R.; Murray, C.; Jhoti, H. A 'Rule Of Three' for Fragment-Based Lead Discovery? *Drug Discov. Today* **2003**, *8*, 876–877.

55. Jhoti, H.; Williams, G.; Rees, D. C.; Murray, C. W. The 'Rule of Three' for Fragment-Based Drug Discovery: Where Are We Now? *Nat. Rev. Drug Discov.* **2013**, *12*, 644–645.

56. Morley, A. D.; Pugliese, A.; Birchall, K.; Bower, J.; Brennan, P.; Brown, N.; Chapman, T.; Drysdale, M.; Gilbert, I. H.; Hoelder, S.; Jordan, A.; Ley, S. V.; Merritt, A.; Miller, D.; Swarbrick, M. E.; Wyatt, P. G. Fragment-Based Hit Identification: Thinking in 3D. *Drug Discov. Today* **2013**, *18*, 1221–1227.

57. Hall, R. J.; Mortenson, P. N.; Murray, C. W. Efficient Exploration of Chemical Space by Fragment-Based Screening. *Prog. Biophys. Mol. Biol.* **2014**, *116*, 82–91.

58. Hann, M. M.; Leach, A. R.; Harper, G. Molecular Complexity and Its Impact on the Probability of Finding Leads for Drug Discovery. *J. Chem. Inf. Comput. Sci.* **2001**, *41*, 856–864.

59. Reymond, J.-L. The Chemical Space Project. *Acc. Chem. Res.* **2015**, *48*, 722–730.

60. Ertl, P. Cheminformatics Analysis of Organic Substituents: Identification of the Most Common Substituents, Calculation of Substituent Properties, and Automatic Identification of Drug-Like Bioisosteric Groups. *J. Chem. Inf. Comput. Sci.* **2003**, *43*, 374–380.

61. Erlanson, D. A.; Fesik, S. W.; Hubbard, R. E.; Jahnke, W.; Jhoti, H. Twenty Years on: The Impact of Fragments on Drug Discovery. *Nat. Rev. Drug Discov.* **2016**, *605*–619.

62. Chen, I.-J.; Hubbard, R. E. Lessons for Fragment Library Design: Analysis of Output From Multiple Screening Campaigns. *J. Comput.-Aided Mol. Des.* **2009**, *23*, 603–620.

63. Lau, W. F.; Withka, J. M.; Hepworth, D.; Magee, T. V.; Du, Y. J.; Bakken, G. A.; Miller, M. D.; Hendsch, Z. S.; Thanabal, V.; Kolodziej, S. A.; Xing, L.; Hu, Q.; Narasimhan, L. S.; Love, R.; Charlton, M. E.; Hughes, S.; van Hoorn, W. P.; Mills, J. E. Design of a Multi-Purpose Fragment Screening Library Using Molecular Complexity and Orthogonal Diversity Metrics. *J. Comput.-Aided Mol. Des.* **2011**, *25*, 621–636.

64. Liu, T.; Naderi, M.; Alvin, C.; Mukhopadhyay, S.; Brylinski, M. Break Down in Order to Build Up: Decomposing Small Molecules for Fragment-Based Drug Design with eMolFrag. *J. Chem. Inf. Model.* **2017**, *57*, 627–631.

65. Ray, P. C.; Kiczun, M.; Huggett, M.; Lim, A.; Prati, F.; Gilbert, I. H.; Wyatt, P. G. Fragment Library Design, Synthesis and Expansion: Nurturing a Synthesis and Training Platform. *Drug Discov. Today* **2017**, *22*, 43–56.

66. Köster, H.; Craan, T.; Brass, S.; Herhaus, C.; Zentgraf, M.; Neumann, L.; Heine, A.; Klebe, G. Small Non-Rule of 3 Compatible Fragment Library Provides High Hit Rate of Endothiapepsin Crystal Structures with Various Fragment Chemotypes. *J. Med. Chem.* **2011**, *54*, 7784–7796.

67. Keseru, G. M.; Erlanson, D. A.; Ferenczy, G. G.; Hann, M. M.; Murray, C. W.; Pickett, S. D. Design Principles for Fragment Libraries: Maximizing the Value of Learnings From Pharma Fragment-Based Drug Discovery (FBDD) Programs for Use in Academia. *J. Med. Chem.* **2016**, *59*, 8189–8206.

68. Davis, B. J.; Erlanson, D. A. Learning From Our Mistakes: The 'Unknown Knowns' in Fragment Screening. *Bioorg. Med. Chem. Lett.* **2013**, *23*, 2844–2852.

69. Gao, H.; Shanmugasundaram, V.; Lee, P. Estimation of Aqueous Solubility of Organic Compounds with QSPR Approach. *Pharm. Res.* **2002**, *19*, 497–503.

70. Alsenz, J.; Kansy, M. High Throughput Solubility Measurement in Drug Discovery and Development. *Adv. Drug Delivery Rev.* **2007**, *59*, 546–567.

71. Seidler, J.; McGovern, S. L.; Doman, T. N.; Shoiket, B. K. Identification and Prediction of Promiscuous Aggregating Inhibitors Among Known Drugs. *J. Med. Chem.* **2003**, *46*, 4477–4486.

72. Siegal, G.; Eiso, A. B.; Schultz, J. Integration of Fragment Screening and Library Design. *Drug Discov. Today* **2007**, *12*, 1032–1039.

73. Schuffenhauer, A.; Ruedisser, S.; Marzinzik, A. L.; Jahnke, W.; Blommers, M.; Selzer, P.; Jacoby, E. Library Design for Fragment Based Screening. *Curr. Top. Med. Chem.* **2005**, *5*, 751–762.

74. Wall, I. D.; Hann, M. M.; Leach, A. R.; Pickett, S. D. Current Status and Future Direction of Fragment-Based Drug Discovery: A Computational Chemistry Perspective. In *Fragment-Based Drug Discovery*; Howard, S., Abell, C., Eds.; London: RSC, 2015; pp 73–100.

75. Lewell, X. Q.; Judd, D. B.; Watson, S. P.; Hann, M. M. RECAP Retrosynthetic Combinatorial Analysis Procedure: A Powerful New Technique for Identifying Privileged Molecular Fragments with Useful Applications in Combinatorial Chemistry. *J. Chem. Inf. Comput. Sci.* **1998**, *38*, 511–522.

76. Murray, C. W.; Rees, D. C. Opportunity Knocks: Organic Chemistry for Fragment-Based Drug Discovery (FBDD). *Angew. Chem., Int. Ed.* **2016**, *55*, 488–492.

77. Dandapani, S.; Rosse, G.; Southall, N.; Salvino, J. M.; Thomas, C. J. Selecting, Acquiring, and Using Small Molecule Libraries for High-Throughput Screening. *Curr. Protoc. Chem. Biol.* **2012**, *4*, 177–191.

78. Mok, N. Y.; Brenk, R.; Brown, N. Increasing the Coverage of Medicinal Chemistry-Relevant Space in Commercial Fragments Screening. *J. Chem. Inf. Model.* **2014**, *54*, 79–85.

79. Austin, C. P.; Brady, L. S.; Insel, T. R.; Collins, F. S. NIH Molecular Libraries Initiative. *Science* **2004**, *306*, 1138–1139.

80. Schreiber, S. L.; Kotz, J. D.; Li, M.; Aube, J.; Austin, C. P.; Reed, J. C.; Rosen, H.; White, E. L.; Sklar, L. A.; Lindsley, C. W.; Alexander, B. R.; Bittker, J. A.; Clemons, P. A.; de Souza, A.; Foley, M. A.; Palmer, M.; Shamji, A. F.; Wawer, M. J.; McManus, O.; Wu, M.; Zou, B.; Yu, H.; Golden, J. E.; Schoenen, F. J.; Simeonov, A.; Jadhav, A.; Jackson, M. R.; Pinkerton, A. B.; Chung, T. D. Y.; Griffin, P. R.; Cravatt, B. F.; Hodder, P. S.; Roush, W. R.; Roberts, E.; Chung, D.-H.; Jonsson, C. B.; Noah, J. W.; Severson, W. E.; Ananthan, S.; Edwards, B.; Oprea, T. I.; Conn, P. J.; Hopkins, C. R.; Wood, M. R.; Stauffer, S. R.; Emmitte, K. A.; NIH Molecular Libraries Project Team. Advancing Biological Understanding and Therapeutics Discovery with Small-Molecule Probes. *Cell* **2015**, *161*, 1252–1265.

81. Mullard, A. European Lead Factory Opens for Business. *Nat. Rev. Drug Discov.* **2013**, *12*, 173–175.

82. Kingwell, K. European Lead Factory Hits its Stride. *Nat. Rev. Drug Discov.* **2016**, *15*, 221–222.

83. Mayer, M. P.; Bukau, B. Hsp70 Chaperones: Cellular Functions and Molecular Mechanisms. *Cell. Mol. Life Sci.* **2005**, *62*, 670–684.

84. Pearl, L. H.; Prodromou, C. Structure and Mechanism of the Hsp90 Molecular Chaperone Machinery. *Annu. Rev. Biochem.* **2006**, *75*, 271–294.

85. Ciechanover, A.; Kwon, Y. T. Protein Quality Control by Molecular Chaperones in Neurodegeneration. *Front. Neurosci.* **2017**, *11*, 185.

86. Chen, B.; Retzlaff, M.; Roos, T.; Frydman, J. Cellular Strategies of Protein Quality Control. *Cold Spring Harb. Perspect. Biol.* **2011**, *3*, a004374.

87. Bukau, B.; Weissman, J.; Horwich, A. Molecular Chaperones and Protein Quality Control. *Cell* **2006**, *125*, 443–451.

88. Kim, R.; Lai, L.; Lee, H. H.; Cheong, G. W.; Kim, K. K.; Wu, Z.; Yokota, H.; Marqusee, S.; Kim, S. H. On the Mechanism of Chaperone Activity of the Small Heat-Shock Protein of *Methanococcus jannaschii. Proc. Natl. Acad. Sci. U.S.A.* **2003**, *100*, 8151–8155.

89. Haslbeck, M.; Miess, A.; Stromer, T.; Walter, S.; Buchner, J. Disassembling Protein Aggregates in the Yeast Cytosol: the Cooperation of HSP26 with Ssa1 and Hsp104. *J. Biol. Chem.* **2005**, *280*, 23861–23868.

90. Slepenkov, S. V.; Witt, S. N. The Unfolding Story of the *Escherichia coli* Hsp70 DnaK: is DnaK a Holdase or an Unfoldase? *Mol. Microbiol.* **2002**, *45*, 1197–1206.

91. Sharma, S. K.; Christen, P.; Goloubinoff, P. Disaggregating Chaperones: an Unfolding Story. *Curr. Protein Pept. Sci.* **2009**, *10*, 432–446.

92. De Los Rios, P.; Ben-Zvi, A.; Slutsky, O.; Azem, A.; Goloubinoff, P. Hsp70 Chaperones Accelerate Protein Translocation and the Unfolding of Stable Protein Aggregates by Entropic Pulling. *Proc. Natl. Acad. Sci. U.S.A.* **2006**, *103*, 6166–6171.

93. Uversky, V. N. Flexible Nets Of Malleable Guardians: Intrinsically Disordered Chaperones in Neurodegenerative Diseases. *Chem. Rev.* **2011**, *111*, 1134–1166.

94. Csermely, P.; Schnaider, T.; Soti, C.; Prohaszka, Z.; Nardai, G. The 90 kDa Molecular Chaperone Family: Structure, Function and Clinical Applications. A Comprehensive Review. *Pharmacol. Ther.* **1998**, *79*, 129–168.

95. Dutta, R.; Inouye, M. GHKL, an Emergent ATPase/Kinase Superfamily. *Trends Biochem. Sci.* **2000**, *25*, 24–28.

96. Hickey, E.; Brandon, S. E.; Smale, G.; Lloyd, D.; Weber, L. A. Sequence and Regulation of a Gene Encoding a Human 89-Kilodalton Heat Shock Protein. *Mol. Cell. Biol.* **1989**, *9*, 2615–2626.

97. Sreedhar, A. S.; Kalmar, E.; Csermely, P.; Shen, Y. F. Hsp90 Isoforms: Functions, Expression and Clinical Importance. *FEBS Lett.* **2004**, *562*, 11–15.

98. Grammatikakis, N.; Vultur, A.; Ramana, C. V.; Siganou, A.; Schweinfest, C. W.; Watson, D. K.; Raptis, L. The Role of Hsp90N, a New Member of the Hsp90 Family, in Signal Transduction and Neoplastic Transformation. *J. Biol. Chem.* **2002**, *277*, 8312–8320.

99. Pearl, L. H.; Prodromou, C.; Workman, P. The Hsp90 Molecular Chaperone: An Open and Shut Case for Treatment. *Biochem. J.* **2008**, *410*, 439–453.

100. Koren, J., III; Jinwal, U. K.; Lee, D. C.; Jones, J. R.; Shults, C. L.; Johnson, A. G.; Anderson, L. J.; Dickey, C. A. Chaperone Signalling Complexes in Alzheimer's Disease. *J. Cell. Mol. Med.* **2009**, *13*, 619–630.

101. Schneider, C.; Sepp-Lorenzino, L.; Nimmesgern, E.; Ouerfelli, O.; Danishefsky, S.; Rosen, N.; Hartl, F. U. Pharmacologic Shifting of a Balance Between Protein Refolding and Degradation Mediated by Hsp90. *Proc. Natl. Acad. Sci. U.S.A.* **1996**, *93*, 14536–14541.

102. http://www.picard.ch/downloads/Hsp90interactors.pdf.

103. Echeverria, P. C.; Bernthaler, A.; Dupuis, P.; Mayer, B.; Picard, D. An Interaction Network Predicted from Public Data as a Discovery Tool: Application to the Hsp90 Molecular Chaperone Machine. *PLoS ONE* **2011**, *6*, e26044.

104. Roe, S. M.; Ali, M. M. U.; Meyer, P.; Vaughan, C. K.; Panaretou, B.; Piper, P. W.; Prodromou, C.; Pearl, L. H. The Mechanism of Hsp90 Regulation by the Protein Kinase-Specific cochaperone p50^{cdc37}. *Cell* **2004**, *116*, 87–98.

105. Picard, D. Heat-Shock Protein 90, a Chaperone for Folding and Regulation. *Cell. Mol. Life Sci.* **2002**, *59*, 1640–1648.

106. Zou, J.; Guo, Y.; Guettouche, T.; Smith, D. F.; Voellmy, R. Repression of Heat Shock Transcription Factor HSF1 Activation by HSP90 (HSP90 complex) That Forms a Stress-Sensitive Complex with HSF1. *Cell* **1998**, *94*, 471–480.

107. Evans, C. G.; Chang, L.; Gestwicki, J. E. Heat Shock Protein 70 (Hsp70) as an Emerging Drug Target. *J. Med. Chem.* **2010**, *53*, 4585–4602.

108. Wegele, H.; Muller, L.; Buchner, J. Hsp70 and Hsp90 - a Relay Team for Protein Folding. *Rev. Physiol. Biochem. Pharmacol.* **2004**, *151*, 1–44.

109. Theodoraki, M. A.; Caplan, A. J. Quality Control and Fate Determination of Hsp90 Client Proteins. *Biochim. Biophys. Acta* **2012**, *1823*, 683–688.

110. Travers, J.; Sharp, S.; Workman, P. HSP90 Inhibition: Two-Pronged Exploitation of Cancer Dependencies. *Drug Discov. Today* **2012**, *17*, 242–252.

111. Sőti, C.; Nagy, E.; Giricz, Z.; Vígh, L.; Csermely, P.; Ferdinandy, P. Heat Shock Proteins as Emerging Therapeutic Targets. *Br. J. Pharmacol.* **2005**, *146*, 769–780.

112. Luo, W.; Sun, W.; Taldone, T.; Rodina, A.; Chiosis, G. Heat Shock Protein 90 in Neurodegenerative Diseases. *Mol. Neurodeg.* **2010**, *5*, 24.

113. NTC00003969. http://www.clinicaltrials.gov/ct2/show/NCT00003969?term=17-aag&rank=34.

114. Jhaveri, K.; Taldone, T.; Modi, S.; Chiosis, G. Advances in the Clinical Development of Heat Shock Protein 90 (Hsp90) Inhibitors in Cancers. *Biochim. Biophys. Acta* **1823**, *2012*, 742–755.

115. Tatokoro, M.; Koga, F.; Yoshida, S.; Kihara, K. Heat Shock Protein 90 Targeting Therapy: State of the Art and Future Perspective. *EXCLI J.* **2015**, *14*, 48–58.

116. DeBoer, C.; Meulman, P. A.; Wnuk, R. J.; Peterson, D. H. Geldanamycin, a New Antibiotic. *J. Antibiot.* **1970**, *23*, 442–447.

117. Schulte, T. W.; Akinaga, S.; Murakata, T.; Agatsuma, T.; Sugimoto, S.; Nakano, H.; Lee, Y. S.; Simen, B. B.; Argon, Y.; Felts, S.; Toft, D. O.; Neckers, L. M.; Sharma, S. V. Interaction of Radicicol with Members of the Heat Shock Protein 90 Family of Molecular Chaperones. *Mol. Endocrinol.* **1999**, *13*, 1435–1448.

118. Chiosis, G.; Timaul, M. N.; Lucas, B.; Munster, P. N.; Zheng, F. F.; Sepp-Lorenzino, L.; Rosen, N. A Small Molecule Designed to Bind to the Adenine Nucleotide Pocket of Hsp90 Causes Her2 Degradation and the Growth Arrest and Differentiation of Breast Cancer Cells. *Chem. Biol.* **2001,** *8,* 289–299.

119. Cheung, K.-M.J.; Matthews, T. P.; James, K.; Rowlands, M. G.; Boxall, K. J.; Sharp, S. Y.; Maloney, A.; Roe, S. M.; Prodromou, C.; Pearl, L. H.; Aherne, G. W.; McDonald, E.; Workman, P. The Identification, Synthesis, Protein Crystal Structure and In Vitro Biochemical Evaluation of a New 3,4-Diarylpyrazole Class of Hsp90 Inhibitors. *Bioorg. Med. Chem. Lett.* **2005,** 3338–3343.

120. Du, Y.; Moulick, K.; Rodina, A.; Aguirre, J.; Felts, S.; Dingledine, R.; Fu, H.; Chiosis, G. High-Throughput Screening Fluorescence Polarization Assay for Tumor-Specific Hsp90. *J. Biomol. Scr.* **2007,** *12,* 915–924.

121. Stebbins, C. E.; Russo, A. A.; Schneider, C.; Rosen, N.; Hartl, F. U.; Pavletich, N. P. Crystal Structure of an Hsp90–Geldanamycin Complex: Targeting of a Protein Chaperone by an Antitumor Agent. *Cell* **1997,** *89,* 239–250.

122. Roe, S. M.; Prodromou, C.; O'Brien, R.; Ladbury, J. E.; Piper, P. W.; Pearl, L. H. Structural Basis for Inhibition of the Hsp90 Molecular Chaperone by the Antitumor Antibiotics Radicicol and Geldanamycin. *J. Med. Chem.* **1999,** *42,* 260–266.

123. Obermann, W. M. J.; Sondermann, H.; Russo, A. A.; Pavletich, N. P.; Hartl, F. U. In Vivo Function of Hsp90 is Dependent on ATP Binding and ATP Hydrolysis. *J. Cell Biol.* **1998,** *143,* 901–910.

124. Joachimiak, A. Capturing the Misfolds: Chaperone-Peptide-Binding Motifs. *Nat. Struct. Biol.* **1997,** *4,* 430–434.

125. Salek, R. M.; Williams, M. A.; Prodromou, C.; Pearl, L. H.; Ladbury, J. E. Letter to the Editor: Backbone Resonance Assignments of the 25kD N-Terminal ATPase Domain from the Hsp90 Chaperone. *J. Biomol. NMR* **2002,** *23,* 327–328.

126. Dehner, A.; Furrer, J.; Richter, K.; Schuster, I.; Buchner, J.; Kessler, H. NMR Chemical Shift Perturbation Study of the N-Terminal Domain of Hsp90 Upon Binding of ADP, AMP-PNP, Geldanamycin, and Radicicol. *ChemBioChem* **2003,** *4,* 870–877.

127. Wright, L.; Barril, X.; Dymock, B.; Sheridan, L.; Surgenor, A.; Beswick, M.; Drysdale, M.; Collier, A.; Massey, A.; Davies, N.; Fink, A.; Fromont, C.; Aherne, W.; Boxall, K.; Sharp, S.; Workman, P.; Hubbard, R. E. Structure-Activity Relationships in Purine-Based Inhibitor Binding to HSP90 Isoforms. *Chem. Biol.* **2004,** *11,* 775–785.

128. Chiosis, G.; Lucas, B.; Shtil, A.; Huezo, H.; Rosen, N. Development of a Purine-Scaffold Novel Class of Hsp90 Binders That Inhibit the Proliferation of Cancer Cells and Induce the Degradation of Her2 Tyrosine Kinase. *Bioorg. Med. Chem.* **2002,** *10,* 3555–3564.

129. Ambati, S. R.; Lopes, E. C.; Kosugi, K.; Mony, U.; Zehir, A.; Shah, S. K.; Taldone, T.; Moreira, A. L.; Meyers, P. A.; Chiosis, G.; Moore, M. A. Pre-Clinical Efficacy of PU-H71, a Novel HSP90 Inhibitor, Alone and in Combination with Bortezomib in Ewing Sarcoma. *Mol. Oncol.* **2014,** *8,* 323–336.

130. Barril, X.; Morley, S. D. Unveiling the Full Potential of Flexible Receptor Docking Using Multiple Crystallographic Structures. *J. Med. Chem.* **2005,** *48,* 4432–4443.

131. Roughley, S.; Wright, L.; Brough, P.; Massey, A.; Hubbard, R. E. Hsp90 Inhibitors and Drugs From Fragment and Virtual Screening. *Top. Curr. Chem.* **2012,** *317,* 61–82.

132. Baurin, N.; Baker, R.; Richardson, C.; Chen, I.; Foloppe, N.; Potter, A.; Jordan, A.; Roughley, S.; Parratt, M.; Greaney, P.; Morley, D.; Hubbard, R. E. Drug-Like Annotation and Duplicate Analysis of a 23-Supplier Chemical Database Totalling 2.7 Million Compounds. *J. Chem. Inf. Compu. Sci.* **2004,** *44,* 643–651.

133. Weininger, D. SMILES 1. Introduction and Encoding Rules. *J. Chem. Inf. Comput. Sci.* **1988**, *28*, 31–36.
134. Huuskonen, J. Estimation of Aqueous Solubility for a Diverse Set of Organic Compounds Based on Molecular Topology. *J. Chem. Inf. Comput. Sci.* **2000**, *40*, 773–777.
135. Egan, W. J.; Lauri, G. Prediction of Intestinal Permeability. *Adv. Drug Deliv. Rev.* **2002**, *54*, 273–289.
136. Aherne, W.; Maloney, A.; Prodromou, C.; Rowlands, M. G.; Hardcastle, A.; Boxall, K.; Clarke, P.; Walton, M. I.; Pearl, L.; Workman, P. Assays for HSP90 and Inhibitors. *Methods Mol. Med.* **2003**, *85*, 149–161.
137. Howes, R.; Barril, X.; Dymock, B. W.; Grant, K.; Northfield, C. J.; Robertson, A. G.; Surgenor, A.; Wayne, J.; Wright, L.; James, K.; Matthews, T.; Cheung, K. M.; McDonald, E.; Workman, P.; Drysdale, M. J. A Fluorescence Polarization Assay for Inhibitors of Hsp90. *Anal. Biochem.* **2006**, *350*, 202–213.
138. Baurin, N.; Aboul-Ela, F.; Barril, X.; Davis, B.; Drysdale, M.; Dymock, B.; Finch, H.; Fromont, C.; Richardson, C.; Simmonite, H.; Hubbard, R. E. Design and Characterization of Libraries of Molecular Fragments for Use in NMR Screening Against Protein Targets. *J. Chem. Inf. Comput. Sci.* **2004**, *44*, 2157–2166.
139. http://www.acdlabs.com/resources/knowledgebase/samples_dbs/.
140. Meijer, L.; Thunnissen, A. M.; White, A. W.; Garnier, M.; Nikolic, M.; Tsai, L. H.; Walter, J.; Cleverley, K. E.; Salinas, P. C.; Wu, Y. Z.; Biernat, J.; Mandelkow, E. M.; Kim, S. H.; Pettit, G. R. Inhibition of Cyclin-Dependent Kinases, GSK-3beta and CK1 by Hymenialdisine, a Marine Sponge Constituent. *Chem. Biol.* **2000**, *7*, 51–63.
141. Cala, O.; Krimm, I. Ligand-Orientation Based Fragment Selection in STD NMR Screening. *J. Med. Chem.* **2015**, *58*, 8739–8742.
142. Dalvit, C.; Fogliatto, G.; Stewart, A.; Veronesi, M.; Stockman, B. WaterLOGSY as a Method for Primary NMR Screening: Practical Aspects and Range of Applicability. *J. Biomol. NMR* **2001**, *21*, 349–359.
143. Vallurupalli, P.; Bouvignies, G.; Kay, L. E. Increasing the Exchange Time-Scale That Can Be Probed by CPMG Relaxation Dispersion NMR. *J. Phys. Chem. B.* **2011**, *115*, 14891–14900.
144. Hubbard, R. E.; Davis, B.; Chen, I.; Drysdale, M. J. The SeeDs Approach: Integrating Fragments into Drug Discovery. *Curr. Top. Med. Chem.* **2007**, *7*, 1568–1581.
145. Rowlands, M. G.; Newbatt, Y. M.; Prodromou, C.; Pearl, L. H.; Workman, P.; Aherne, W. High-Throughput Screening Assay for Inhibitors of Heat-Shock Protein 90 ATPase Activity. *Anal. Biochem.* **2004**, *327*, 176–183.
146. Sharp, S. Y.; Boxall, K.; Rowlands, M.; Prodromou, C.; Roe, S. M.; Maloney, A.; Powers, M.; Clarke, P. A.; Box, G.; Sanderson, S.; Patterson, L.; Matthews, T. P.; Cheung, K. M.; Ball, K.; Hayes, A.; Raynaud, F.; Marais, R.; Pearl, L.; Eccles, S.; Aherne, W.; McDonald, E.; Workman, P. In Vitro Biological Characterization of a Novel, Synthetic diaryl Pyrazole Resorcinol Class of Heat Shock Protein 90 Inhibitors. *Cancer Res.* **2007**, *67*, 2206–2216.
147. Smith, N. F.; Hayes, A.; James, K.; Nutley, B. P.; McDonald, E.; Henley, A.; Dymock, B.; Drysdale, M. J.; Raynaud, F. I.; Workman, P. Preclinical Pharmacokinetics and Metabolism of a Novel Diaryl Pyrazole Resorcinol Series of Heat Shock Protein 90 Inhibitors. *Mol. Canc. Ther.* **2006**, *5*, 1628–1637.
148. Barril, X.; Brough, P.; Drysdale, M.; Hubbard, R. E.; Massey, A.; Surgenor, A.; Wright, L. Structure-Based Discovery of a New Class of Hsp90 Inhibitors. *Bioorg. Med. Chem. Lett.* **2005**, *15*, 5187–5191.

149. Brough, P. A.; Barril, X.; Borgognoni, J.; Chene, P.; Davies, N. G. M.; Davis, B.; Drysdale, M. J.; Dymock, B.; Eccles, S. A.; Garcia-Echeverria, C.; Fromont, C.; Hayes, A.; Hubbard, R. E.; Jordan, A. M.; Rugaard Jensen, M.; Massey, A.; Merrett, A.; Padfield, A.; Parsons, R.; Radimerski, T.; Raynaud, F. I.; Robertson, A.; Roughley, S. D.; Schoepfer, J.; Simmonite, H.; Sharp, S. Y.; Surgenor, A.; Valenti, M.; Walls, S.; Webb, P.; Wood, M.; Workman, P.; Wright, L. Combining Hit Identification Strategies: Fragment-Based and In Silico Approaches to Orally Active 2-Aminothieno[2,3-d]Pyrimidine Inhibitors of the Hsp90 Molecular Chaperone. *J. Med. Chem.* **2009**, *52*, 4794–4809.

150. Massey, A. J.; Schoepfer, J.; Brough, P. A.; Brueggen, J.; Chène, P.; Drysdale, M. J.; Pfaar, U.; Radimerski, T.; Ruetz, S.; Schweitzer, A.; Wood, M.; Garcia-Echeverria, C.; Rugaard Jensen, M. Preclinical Antitumor Activity of the Orally Available Heat Shock Protein 90 Inhibitor NVP-BEP800. *Mol. Cancer Ther.* **2010**, *9*, 906–919.

151. Davies, N. G. M.; Browne, H.; Davis, B.; Drysdale, M. J.; Foloppe, N.; Geoffrey, S.; Gibbons, B.; Hart, T.; Hubbard, R.; Rugaard Jensen, M.; Mansell, H.; Massey, A.; Matassova, N.; Moore, J. D.; Murray, J.; Pratt, R.; Ray, S.; Robertson, A.; Roughley, S. D.; Schoepfer, J.; Scriven, K.; Simmonite, H.; Stokes, S.; Surgenor, A.; Webb, P.; Wood, M.; Wright, L.; Brough, P. Targeting Conserved Water Molecules: Design of 4-aryl-5-cyanopyrrolo[2,3-d]Pyrimidine Hsp90 Inhibitors Using Fragment-Based Screening and Structure-Based Optimization. *Bioorg. Med. Chem.* **2012**, *20*, 6770–6789.

152. Dymock, B. W.; Barril, X.; Brough, P. A.; Cansfield, J. E.; Massey, A.; McDonald, E.; Hubbard, R. E.; Surgenor, A.; Roughley, S.; Webb, P.; Workman, P.; Wright, L.; Drysdale, M. Novel, Potent Small-Molecule Inhibitors of the Molecular Chaperone Hsp90 Discovered Through Structure-Based Design. *J. Med. Chem.* **2005**, *48*, 4212–4215.

153. Brough, P. A.; Barril, X.; Beswick, M.; Dymock, B. W.; Drysdale, M. J.; Wright, L.; Grant, K.; Massey, A.; Surgenor, A.; Workman, P. 3-(5-Chloro-2,4-dihydroxyphenyl)-Pyrazole-4-Carboxamides as Inhibitors of the Hsp90 Molecular Chaperone. *Bioorg. Med. Chem. Lett.* **2005**, *15*, 5197–5201.

154. Barril, X.; Beswick, M. C.; Collier, A.; Drysdale, M. J.; Dymock, B. W.; Fink, A.; Grant, K.; Howes, R.; Jordan, A. M.; Massey, A.; Surgenor, A.; Wayne, J.; Workman, P.; Wright, L. 4-Amino Derivatives of the Hsp90 Inhibitor CCT018159. *Bioorg. Med. Chem. Lett.* **2006**, *16*, 2543–2548.

155. Brough, P. A.; Aherne, W.; Barril, X.; Borgognoni, J.; Boxall, K.; Cansfield, J. E.; Cheung, K.-M.J.; Collins, I.; Davies, N. G. M.; Drysdale, M. J.; Dymock, B.; Eccles, S. A.; Finch, H.; Fink, A.; Hayes, A.; Howes, R.; Hubbard, R. E.; James, K.; Jordan, A. M.; Lockie, A.; Martins, V.; Massey, A.; Matthews, T. P.; McDonald, E.; Northfield, C. J.; Pearl, L. H.; Prodromou, C.; Ray, S.; Raynaud, F. I.; Roughley, S. D.; Sharp, S. Y.; Surgenor, A.; Walmsley, D. L.; Webb, P.; Wood, M.; Workman, P.; Wright, L. 4,5-Diarylisoxazole Hsp90 Chaperone Inhibitors: Potential Therapeutic Agents for the Treatment of Cancer. *J. Med. Chem.* **2008**, *51*, 196–218.

156. Hu, Y.; Stumpfe, D.; Bajorath, J. Recent Advances in Scaffold Hopping. *J. Med. Chem.* **2017**, *60*, 1238–1246.

157. Sharp, S. Y.; Prodromou, C.; Boxall, K.; Powers, M. V.; Holmes, J. L.; Box, G.; Matthews, T. P.; Cheung, K. M.; Kalusa, A.; James, K.; Hayes, A.; Hardcastle, A.; Dymock, B.; Brough, P. A.; Barril, X.; Cansfield, J. E.; Wright, L.; Surgenor, A.; Foloppe, N.; Hubbard, R. E.; Aherne, W.; Pearl, L.; Jones, K.; McDonald, E.; Raynaud, F.; Eccles, S.; Drysdale, M.; Workman, P. Inhibition of the Heat Shock Protein 90 Molecular Chaperone in Vitro and in Vivo by Novel, Synthetic, Potent Resorcinylic Pyrazole/Isoxazole Amide Analogues. *Mol. Cancer Ther.* **2007**, *6*, 1198–1211.

158. Eccles, S. A.; Massey, A.; Raynaud, F.; Sharp, S. Y.; Box, G.; Valenti, M.; Patterson, L.; Gowan, S.; Boxall, F.; Aherne, W.; Rowlands, M.; Hayes, A.; Martins, V.; Urban, F.; Boxall, K.; Prodromou, C.; Pearl, L.; James, K.; Matthews, T.; Cheung, K. M.; Kalusa, A.; Jones, K.; McDonald, E.; Barril, X.; Brough, P. A.; Cansfield, J. E.; Dymock, B.; Drysdale, M.; Finch, H.; Howes, R.; Hubbard, R. E.; Surgenor, A.; Webb, P.; Wood, L. M.; Wright, L.; Workman, P. NVP-AUY922: a Novel Synthetic Resorcinylic Isoxazole Amide HSP90 Inhibitor with Potent Activity Against Xenograft Tumor Growth, Angiogenesis and Metastasis. *Cancer Res.* **2007**, *68*, 2850–2860.

159. Johnson, M. L.; Yu, H. A.; Hart, E. M.; Weitner, B. B.; Rademaker, A. W.; Patel, J. D.; Kris, M. G.; Riely, G. J. Phase I/II Study of HSP90 Inhibitor AUY922 and Erlotinib for EGFR-Mutant Lung Cancer with Acquired Resistance to Epidermal Growth Factor Receptor Tyrosine Kinase Inhibitors. *J. Clin. Oncol.* **2015**, *33*, 1666–1673.

160. Seggewiss-Bernhardt, R.; Bargou, R. C.; Tee Goh, Y.; Stewart, A. K.; Spencer, A.; Alegre, A.; Blad, J.; Ottmann, O. G.; Fernandez-Ibarra, C.; Lu, H.; Pain, S.; Akimov, M.; Padmanabhan Iyer, S. Phase 1/1B Trial of the Heat Shock Protein 90 Inhibitor NVP-AUY922 as Monotherapy or in Combination with Bortezomib in Patients with Relapsed or Refractory Multiple Myeloma. *Cancer* **2015**, *121*, 2185–2192.

161. Kong, A.; Rea, D.; Ahmed, S.; Beck, J. T.; López López, R.; Biganzoli, L.; Armstrong, A. C.; Aglietta, M.; Alba, E.; Campone, M.; Hsu Schmitz, S.-F.; Lefebvre, C.; Akimov, M.; Lee, S.-C. Phase 1B/2 Study of the HSP90 Inhibitor AUY922 plus Trastuzumab in Metastatic HER2-Positive Breast Cancer Patients Who Have Progressed on Trastuzumab-Based Regimen. *Oncotarget* **2016**, *7*, 37680–37692.

162. Hendricks, L. E. L.; Dingemans, A.-M.C. Heat Shock Protein Antagonists in Early Stage Clinical Trials for NSCLC. *Exp. Opin. Invest. Drugs* **2017**, *26*, 541–550.

163. http://www.vernalis.com/nce-pipeline/oncology/auy922.

164. Harvey, A. L.; Edrada-Ebel, R.; Quinn, R. J. The Re-Emergence of Natural Products for Drug discovery in the Genomics Era. *Nat. Rev. Drug Discov.* **2015**, *14*, 111–129.

165. Fan, T.-P.; Briggs, J.; Liu, L.; Lu, A.; van der Greef, J.; Xu, A. The Arts and Science of Traditional medicine. Part 1: TCM Today: A Case for Integration. *Science* **2014**, *346*(6216 Suppl), S1–S25.

166. Fleming, A. On the Antibacterial Action of Cultures of a Penicillium, with Special Reference to Their Use in the Isolation of *B. influenzae. Br. J. Exp. Pathol.* **1929**, *10*, 226–236.

167. Chain, E.; Florey, H. W.; Gardner, A. D.; Heatley, N. G.; Jennings, M. A.; Orr-Ewing, J.; Sanders, A. G. Penicillin as a Chemotherapeutic Agent. *Lancet* **1940**, *236*, 226–228.

168. Abraham, E. P.; Chain, E.; Fletcher, C. M.; Florey, H. W.; Gardner, A. D.; Heatley, N. G.; Jennings, M. A. Further Observations on Penicillin. *Lancet* **1941**, *238*, 177–188.

169. Harvey, A. L. Natural Products as a Screening Resource. *Curr. Opin. Chem. Biol.* **2005**, *11*, 480–484.

170. Koehn, F. E.; Carter, G. T. The Evolving Role of Natural Products in Drug Discovery. *Nat. Rev. Drug Discov.* **2005**, *4*, 206–220.

171. Li, J.W.-H.; Vederas, J. C. Drug Discovery and Natural products: End of an Era or an Endless Frontier? *Science* **2009**, *325*, 161–165.

172. Wessjohann, L. A. Synthesis of Natural-Product-Based Compound Libraries. *Curr. Opin. Chem. Biol.* **2000**, *4*, 303–309.

173. Laraia, L.; Waldmann, H. Natural Product Inspired Compound Collections: Evolutionary Principle, Chemical Synthesis, Phenotypic Screening, and Target Identification. *Drug Discov. Today: Technologies* **2017**, *23*, 75–82.

174. Barnes, E. C.; Kumar, R.; Davis, R. A. The Use of Isolated Natural Products as Scaffolds for the Generation of Chemically Diverse Screening Libraries for Drug Discovery. *Nat. Prod. Rep.* **2016**, *33*, 372–381.

175. Xiao, X.-Y.; Parandoosh, Z.; Nova, M. P. Design and Synthesis of a Taxoid Library Using Radiofrequency Encoded Combinatorial Chemistry. *J. Org. Chem.* **1997**, *62*, 6029–6033.

176. Nicolaou, K. C.; Winssinger, N.; Vourloumis, D.; Ohshima, T.; Kim, S.; Pfefferkorn, J.; Xu, J. Y.; Li, T. Solid and Solution Phase Synthesis and Biological Evaluation of Combinatorial Sarcodictyin Libraries. *J. Am. Chem. Soc.* **1998**, *120*, 10814–10826.

177. Wender, P. A.; Verma, V. A.; Paxton, T. J.; Pillow, T. H. Function-Oriented Synthesis, Step Economy, and Drug Design. *Acc. Chem. Res.* **2008**, *41*, 40–49.

178. Nicolaou, K. C.; Pfefferkorn, J. A.; Roecker, A. J.; Cao, G.-Q.; Barluenga, S.; Mitchell, H. J. Natural Product-Like Combinatorial Libraries Based on Privileged Structures. 1. General Principles and Solid-Phase Synthesis of Benzopyrans. *J. Am. Chem. Soc.* **2000**, *122*, 9939–9953.

179. Nicolaou, K. C.; Pfefferkorn, J. A.; Mitchell, H. J.; Roecker, A. J.; Barluenga, S.; Cao, G.-Q.; Affleck, R. L.; Lillig, J. E. Natural Product-Like Combinatorial Libraries Based on Privileged Structures. 2. Construction of a 10 000-Membered Benzopyran Library by Directed Split-and-Pool Chemistry Using NanoKans and Optical Encoding. *J. Am. Chem. Soc.* **2000**, *122*, 9954–9967.

180. Stratton, C. F.; Newman, D. J.; Tan, D. S. Cheminformatic Comparison of Approved Drugs From Natural Product Versus Synthetic Origins. *Bioorg. Med. Chem. Lett.* **2015**, *25*, 4802–4807.

181. Schreiber, S. L. Target-Oriented and Diversity-Oriented Organic Synthesis in Drug Discovery. *Science* **2000**, *287*, 1964–1969.

182. Barjau, J.; Schnakenburg, G.; Waldvogel, S. R. Diversity-Oriented Synthesis of Polycyclic Scaffolds by Modification of an Anodic Product Derived from 2,4-Dimethylphenol. *Angew. Chem. Int. Ed.* **2011**, *50*, 1415–1419.

183. O' Connor, C. J.; Beckmann, H. S. G.; Spring, D. R. Diversity-Oriented Synthesis: Producing Chemical Tools for Dissecting Biology. *Chem. Soc. Rev.* **2012**, *41*, 4444–4456.

184. Wetzel, S.; Bon, R. S.; Kumar, K.; Waldmann, H. Biology-Oriented Synthesis. *Angew. Chem. Int. Ed.* **2011**, *50*, 10800–10826.

185. Liu, W.; Khedkar, V.; Baskar, B.; Schermann, M.; Kumar, K. Branching Cascades: A Concise Synthetic Strategy Targeting Diverse and Complex Molecular Frameworks. *Angew. Chem. Int. Ed.* **2011**, *50*, 6900–6905.

186. Van Hattum, H.; Waldmann, H. Biology-Oriented Synthesis: Harnessing the Power of Evolution. *J. Am. Chem. Soc.* **2014**, *136*, 11853–11859.

187. Wessjohann, L. A.; Ruijter, E.; Garcia-Rivera, D.; Brandt, W. What Can a Chemist Learn from Nature's Macrocycles? – A Brief, Conceptual View. *Mol. Diversity* **2005**, *9*, 171–186.

188. Wiley, P. F.; Gerzon, K.; Flynn, E. H.; Sigal, M. W., Jr.; Weaver, O.; Quarck, U. C.; Chauvette, R. R.; Monahan, R. Erythromycin. X. Structure of Erythromycin. *J. Am. Chem. Soc.* **1957**, *79*, 6062–6070.

189. Hoefler, G.; Bedorf, N.; Steinmetz, H.; Schomburg, D.; Gerth, K.; Reichenbach, H. Epothilone A and B—Novel 16-Membered Macrolides with Cytotoxic Activity: Isolation, Crystal Structure, and Conformation in Solution. *Angew. Chem. Int. Ed.* **1996**, *35*, 1567–1569.

190. Margalith, P.; Beretta, G. Rifomycin. XI. Taxonomic Study on *Streptomyces mediterranei* nov. sp. *Mycopathol. Mycol. Appl.* **1960**, *8*, 321–330.

191. Kino, T.; Hatanaka, H.; Hashimoto, M.; Nishiyama, M.; Goto, T.; Okuhara, M.; Kohsaka, M.; Aoki, H.; Imanaka, H. FK-506, a Novel Immunosuppressant Isolated from a Streptomyces. I. Fermentation, Isolation, and Physico-Chemical and Biological Characteristics. *J. Antibiot.* **1987**, *40*, 1249–1255.

192. Dutcher, J. D. The Discovery and Development of Amphotericin B. *Dis. Chest* **1968**, *54*(Suppl. 1), 296–298.

193. Borel, J. F.; Feurer, C.; Gubler, H. U.; Stähelin, H. Biological Effects of Cyclosporin A: A New Antilymphocytic Agent. *Agents Actions* **1976**, *6*, 468–475.

194. Giordanetto, F.; Kihlberg, J. Macrocyclic Drugs and Clinical Candidates: What Can Medicinal Chemists Learn from Their Properties? *J. Med. Chem.* **2014**, *57*, 278–295.

195. Driggers, E. M.; Hale, S. P.; Lee, J.; Terrett, N. K. The Exploration of Macrocycles for Drug Discovery—An Underexploited Structural Class. *Nat. Rev. Drug Discov.* **2008**, *7*, 608–624.

196. Jones, S.; Thornton, J. M. Principles of Protein-Protein Interactions. *Proc. Natl. Acad. Sci. U.S.A.* **1996**, *93*, 13–20.

197. Petta, I.; Lievens, S.; Libert, C.; Tavernier, J.; De Bosscher, K. Modulation of Protein–Protein Interactions for the Development of Novel Therapeutics. *Mol. Therapy* **2015**, *24*, 707–718.

198. Laraia, L.; McKenzie, G.; Spring, D. R.; Venkitaraman, A. R.; Huggins, D. J. Overcoming Chemical, Biological, and Computational Challenges in the Development of Inhibitors Targeting Protein-Protein Interactions. *Chem. Biol.* **2015**, *22*, 689–703.

199. Bier, D.; Thiel, P.; Briels, J.; Ottmann, C. Stabilization of Protein-Protein Interactions in Chemical Biology and Drug Discovery. *Progr. Biophys. Mol. Biol.* **2015**, *119*, 10–19.

200. Fischer, G.; Rossmann, M.; Hyvönen, M. Alternative Modulation of Protein–Protein Interactions by Small Molecules. *Curr. Opin. Biotechnol.* **2015**, *35*, 78–85.

201. Milroy, L.-G.; Grossmann, T. N.; Hennig, S.; Brunsveld, L.; Ottmann, C. Modulators of Protein-Protein Interactions. *Chem. Rev.* **2014**, *114*, 4695–4748.

202. Dougherty, P. G.; Qian, Z.; Pei, D. Macrocycles as Protein-Protein Interaction Inhibitors. *Biochem. J.* **2017**, *474*, 1109–1125.

203. Villar, E. A.; Beglov, D.; Chennamadhavuni, S.; Porco, J. A., Jr.; Kozakov, D.; Vajda, S.; Whitty, A. How Proteins Bind Macrocycles. *Nat. Chem. Biol.* **2014**, *10*, 723–731.

204. Kozakov, D.; Hall, D. R.; Chuang, G.-Y.; Cencic, R.; Brenke, R.; Grove, L. E.; Beglov, D.; Pelletier, J.; Whitty, A.; Vajda, S. Structural Conservation of Druggable Hot Spots in Protein-Protein Interfaces. *Proc. Natl. Acad. Sci. USA* **2011**, *109*, 13528–13533.

205. Vieth, M.; Siegel, M. G.; Higgs, R. E.; Watson, I. A.; Robertson, D. H.; Savin, K. A.; Durst, G. L.; Hipskind, P. A. Characteristic Physical Properties and Structural Fragments of Marketed Oral Drugs. *J. Med. Chem.* **2004**, *47*, 224–232.

206. Whitty, A.; Zhong, M.; Viarengo, L.; Beglov, D.; Hall, D. R.; Vajda, S. Quantifying the Chameleonic Properties of Macrocycles and Other High-Molecular-Weight Drugs. *Drug Discov. Today* **2016**, *21*, 712–717.

207. Bockus, A. T.; McEwen, C. M.; Lokey, R. S. Form and Function in Cyclic Peptide Natural Products: A Pharmacokinetic Perspective. *Curr. Top. Med. Chem.* **2013**, *13*, 821–836.

208. Troeira Henriques, S.; Huang, Y.-H.; Castanho, M.A.R.B.; Bagatolli, L. A.; Sonza, S.; Tachedjian, G.; Daly, N. L.; Craik, D. J. Phosphatidylethanolamine Binding Is a Conserved Feature of Cyclotide-Membrane Interactions. *J. Biol. Chem.* **2012**, *287*, 33629–33643.

209. Yudin, A. Macrocycles: Lessons from the Distant Past, Recent Developments, and Future Directions. *Chem. Sci.* **2015**, *6*, 30–49.

210. Blankenstein, J.; Zhu, J. Conformation-Directed Macrocyclization Reactions. *Eur. J. Org. Chem.* **2005**, 1949–1964.

211. Meutermans, W. D. F.; Bourne, G. T.; Golding, S. T.; Horton, D. A.; Campitelli, M. R.; Craik, D.; Scanlon, M.; Smythe, M. L. Difficult Macrocyclizations: New Strategies for Synthesizing Highly Strained Cyclic Tetrapeptides. *Org. Lett.* **2003**, *5*, 2711–2714.

212. Marsault, E.; Peterson, M. L. Macrocycles Are Great Cycles: Applications, Opportunities, and Challenges of Synthetic Macrocycles in Drug Discovery. *J. Med. Chem.* **2011**, *54*, 1961–2004.

213. Collins, J. C.; James, K. Emac – A Comparative Index for the Assessment of Macrocycliza-tion Efficiency. *MedChemComm* **2012**, *3*, 1489–1495.

214. Over, B.; Matsson, P.; Tyrchan, C.; Artursson, P.; Doak, B. C.; Foley, M. A.; Hilgendorf, C.; Johnston, S. E.; Lee, M. D., IV; Lewis, R. J.; McCarren, P.; Muncipinto, G.; Norinder, U.; Perry, M. W. D.; Duvall, J. R.; Kihlberg, J. Structural and Conformational Determinants of Macrocycle Cell Permeability. *Nat. Chem. Biol.* **2016**, 1065–1074.

215. Press, B. Optimization of the Caco-2 Permeability Assay to Screen Drug Compounds for Intestinal Absorption and Efflux. *Methods Mol. Biol.* **2011**, *763*, 139–154.

216. Varma, M. V. S.; Sateesh, K.; Panchagnula, R. Functional Role of P-Glycoprotein in Limit-ing Intestinal Absorption of Drugs: Contribution of Passive Permeability to P-Glycoprotein Mediated Efflux Transport. *Mol. Pharmaceutics* **2005**, *2*, 12–21.

217. Mallinson, J.; Collins, I. Macrocycles in New Drug Discovery. *Future Med. Chem.* **2012**, *4*, 1409–1438.

218. Allen, S. E.; Dokholyan, N. V.; Bowers, A. A. Dynamic Docking of Conformationally Con-strained Macrocycles: Methods and Applications. *ACS Chem. Biol.* **2016**, *11*, 10–24.

219. Collins, S.; Bartlett, S.; Nie, F.; Sore, H. F.; Spring, D. R. Diversity-Oriented Synthesis of Mac-rocycle Libraries for Drug Discovery and Chemical Biology. *Synthesis* **2016**, *48*, 1457–1473.

220. Yoo, B.; Shin, S. B. Y.; Huang, M. L.; Kirshenbaum, K. Peptoid Macrocycles: Making the Rounds with Peptidomimetic Oligomers. *Chem. Eur. J.* **2010**, *16*, 5528–5537.

221. Clausen, M. H.; Madsen, C. M. Biologically Active Macrocyclic Compounds – From Natural Products to Diversity-Oriented Synthesis. *Eur. J. Org. Chem.* **2011**, 3107–3125.

222. Terrett, N. Methods for the Synthesis of Macrocycle Libraries for Drug Discovery. *Drug Discov. Today: Technol.* **2010**, e97-e104.

223. Bionda, N.; Cryan, A. L.; Fasan, R. Bioinspired Strategy for the Ribosomal Synthesis of Thioether-Bridged Macrocyclic Peptides in Bacteria. *ACS Chem. Biol.* **2014**, *9*, 2008–2013.

224. Bionda, N.; Fasan, R. Ribosomal Synthesis of Natural-Product-Like Bicyclic Peptides in *Escherichia coli*. *ChemBioChem* **2015**, *16*, 2011–2016.

225. Owens, A. E.; de Paola, I.; Hansen, W. A.; Liu, Y.-W.; Khare, S. D.; Fasan, R. Design and Evolution of a Macrocyclic Peptide Inhibitor of the Sonic Hedgehog/Patched Interaction. *J. Am. Chem. Soc.* **2017**, *139*, 12559–12568.

226. Ohuchi, M.; Murakami, H.; Suga, H. The Flexizyme System: A Highly Flexible tRNA Ami-noacylation Tool for the Translation Apparatus. *Curr. Opin. Chem. Biol.* **2007**, *11*, 537–542.

227. Goto, Y.; Katoh, T.; Suga, H. Flexizymes for Genetic Code Reprogramming. *Nat. Protoc.* **2011**, *6*, 779–790.

228. Roberts, R. W.; Szostak, J. W. RNA-Peptide Fusions for the In Vitro Selection of Peptides and Proteins. *Proc. Natl. Acad. Sci. USA* **1997**, *94*, 12297–12302.

229. Passioura, T.; Suga, H. A RaPID Way to Discover Nonstandard Macrocyclic Peptide Modu-lators of Drug Targets. *Chem. Comm.* **2017**, *53*, 1931–1940.

230. Kawamura, A.; Munzel, M.; Kojima, T.; Yapp, C.; Bhushan, B.; Goto, Y.; Tumber, A.; Katoh, T.; King, O. N. F.; Passioura, T.; Walport, L. J.; Hatch, S. B.; Madden, S.; Muller, S.; Brennan, P. E.; Chowdhury, R.; Hopkinson, R. J.; Suga, H.; Schofield, C. J. Highly Selective Inhibition of Histone Demethylases by De Novo Macrocyclic Peptides. *Nat. Commun.* **2016**, *8*, 14773.

231. Salamon, H.; Klika Skopic, M.; Jung, K.; Bugain, O.; Brunschweiger, A. Chemical Biology Probes from Advanced DNA-Encoded Libraries. *ACS Chem. Biol.* **2016**, *11*, 296–307.

232. Goodnow, R. A., Jr.; Dumelin, C. E.; Keefe, A. D. DNA-Encoded Chemistry: Enabling the Deeper Sampling of Chemical Space. *Nat. Rev. Drug Discov.* **2017**, *16*, 131–147.

233. Connors, W. H.; Hale, S. P.; Terrett, N. K. DNA-Encoded Chemical Libraries of Macro-cycles. *Curr. Opin. Chem. Biol.* **2015**, *26*, 42–47.

234. Belyanskaya, S. L.; Ding, Y.; Callahan, Y. F.; Lazaar, A. L.; Israel, D. I. Discovering Drugs with DNA-Encoded Library Technology: From Concept to Clinic With an Inhibitor of Soluble Epoxide Hydrolase. *ChemMedChem* **2017**, *18*, 837–842.

235. Brenner, S.; Lerner, R. A. Encoded Combinatorial Chemistry. *Proc. Natl Acad. Sci. USA* **1992**, *89*, 5381–5383.

236. MacConnell, A. B.; McEnaney, P. J.; Cavett, V. J.; Paegel, B. M. DNA-Encoded Solid-Phase Synthesis: Encoding Language Design and Complex Oligomer Library Synthesis. *ASC Comb. Sci.* **2015**, *17*, 518–534.

237. Keefe, A. D.; Clark, M. A.; Hupp, C. D.; Litovchick, A.; Zhang, Y. Chemical Ligation Methods for the Tagging of DNA-Encoded Chemical Libraries. *Curr. Opin. Chem. Biol.* **2015**, *26*, 80–88.

238. Wrenn, S. J.; Weisinger, R. M.; Halpin, D. R.; Harbury, P. B. Synthetic Ligands Discovered by In Vitro Selection. *J. Am. Chem. Soc.* **2007**, *129*, 13137–13143.

239. Deng, H.; O'Keefe, H.; Davie, C. P.; Lind, K. E.; Acharya, R. A.; Franklin, G. J.; Larkin, J.; Matico, R.; Neeb, M.; Thompson, M. M.; Lohr, T.; Gross, J. W.; Centrella, P. A.; O'Donovan, G. K.; Bedard, K. L.; van Vloten, K.; Mataruse, S.; Skinner, S. R.; Belyanskaya, S. L.; Carpenter, T. Y.; Shearer, T. W.; Clark, M. A.; Cuozzo, J. W.; Arico-Muendel, C. C.; Morgan, B. A. Discovery of Highly Potent and Selective Small Molecule ADAMTS-5 Inhibitors That Inhibit Human Cartilage Degradation via Encoded Library Technology (ELT). *J. Med. Chem.* **2012**, *55*, 7061–7079.

240. Li, X.; Liu, D. R. DNA-Templated Organic Synthesis: Nature's Strategy for Controlling Chemical Reactivity Applied to Synthetic Molecules. *Angew. Chem., Int. Ed.* **2004**, *43*, 4848–4870.

241. Catalano, M. J.; Price, N. E.; Gates, K. S. Effective Molarity in a Nucleic Acid-Controlled Reaction. *Bioorg. Med. Chem. Lett.* **2016**, *26*, 2627–2630.

242. Seigal, B. A.; Connors, W. H.; Fraley, A.; Borzilleri, R. M.; Carter, P. H.; Emanuel, S. L.; Fargnoli, J.; Kim, K.; Lei, M.; Naglich, J. G.; Pokross, M. E.; Posy, S. L.; Shen, H.; Surti, N.; Talbott, R.; Zhang, Y.; Terrett, N. K. The Discovery of Macrocyclic XIAP Antagonists from a DNA-Programmed Chemistry Library, and Their Optimization to Give Lead Compounds with In Vivo Antitumor Activity. *J. Med. Chem.* **2015**, *58*, 2855–2861.

243. Hansen, M. H.; Blakskjaer, P.; Petersen, L. K.; Hansen, T. H.; Hoejfeldt, J. W.; Gothelf, K. V.; Hansen, N. J. V. A Yoctoliter-Scale DNA Reactor for Small Molecule Evolution. *J. Am. Chem. Soc.* **2009**, *131*, 1322–1327.

244. Petersen, L. K. Novel p38α MAP Kinase Inhibitors Identified from yoctoReactor DNA-Encoded Small Molecule Library. *MedChemComm* **2016**, *7*, 1332–1339.

245. Blakskjaer, P.; Heitner, T.; Hansen, N. J. Fidelity by Design: Yoctoreactor and Binder Trap Enrichment for Small-Molecule DNA-Encoded Libraries and Drug Discovery. *Curr. Opin. Chem. Biol.* **2015**, *26*, 62–71.

246. Scheuermann, J.; Neri, D. Dual Pharmacophore DNA-Encoded Chemical Libraries. *Curr. Opin. Chem. Biol.* **2015**, *26*, 99–103.

247. Bigatti, M.; Dal Corso, A.; Vanetti, S.; Cazzamalli, S.; Rieder, U.; Scheuermann, J.; Neri, D.; Sladojevich, F. Impact of a Central Scaffold on the Binding Affinity of Fragment Pairs Isolated from DNA-Encoded Self-Assembling Chemical Libraries. *ChemMedChem* **2017**, *12*, 1748–1752.

248. Wichert, M.; Krall, N.; Decurtins, W.; Franzini, R. M.; Pretto, F.; Schneider, P.; Neri, D.; Scheuermann, J. Dual-Display of Small Molecules Enables the Discovery of Ligand Pairs and Facilitates Affinity Maturation. *Nat. Chem.* **2015**, *7*, 241–249.

249. Chan, A. I.; McGregor, L. M.; Liu, D. R. Novel Selection Methods for DNA-Encoded Chemical Libraries. *Curr. Opin. Chem. Biol.* **2015**, *26*, 55–61.

250. Decurtins, W.; Wichert, M.; Franzini, R. M.; Buller, F.; Stravs, M. A.; Zhang, Y.; Neri, D.; Scheuermann, J. Automated Screening for Small Organic Ligands Using DNA-Encoded Chemical Libraries. *Nat. Protoc.* **2016**, *11*, 764–780.

251. Mannocci, L.; Zhang, Y.; Scheuermann, J.; Leimbacher, M.; De Bellis, G.; Rizzi, E.; Dumelin, C.; Melkko, S.; Neri, D. High Throughput Sequencing Allows the Identification of Binding Molecules Isolated from DNA-Encoded Chemical Libraries. *Proc. Natl. Acad. Sci. U. S. A.* **2008**, *105*, 17670–17675.

252. Winssinger, N.; Damoiseaux, R.; Tully, D. C.; Geierstanger, B. H.; Burdick, K.; Harris, J. L. PNA-Encoded Protease Substrate Microarrays. *Chem. Biol.* **2004**, *11*, 1351–1360.

253. Zhao, P.; Chen, Z.; Li, Y.; Sun, D.; Gao, Y.; Huang, Y.; Li, X. Selection of DNA-Encoded Small Molecule Libraries Against Unmodified and Non-Immobilized Protein Targets. *Angew. Chem. Int. Ed.* **2014**, *53*, 10056–10059.

254. Franzini, R. M.; Randolph, C. Chemical Space of DNA-Encoded Libraries. *J. Med. Chem.* **2016**, *59*, 6629–6644.

255. Arico-Muendel, C. C. From Haystack to Needle: Finding Value with DNA Encoded Library Technology at GSK. *MedChemComm.* **2016**, *7*, 1898–1909.

256. Liu, K.; Zheng, L.; Liu, Q.; de Vries, J. W.; Gerasimov, J. Y.; Herrmann, A. Nucleic Acid Chemistry in the Organic Phase: From Functionalized Oligonucleotides to DNA Side Chain Polymers. *J. Am. Chem. Soc.* **2014**, *136*, 14255–14262.

257. Perkins, G. L.; Langley, J. Analytical Challenges for DNA-Encoded Library Systems. In *A Handbook for DNA-Encoded Chemistry: Theory and Applications for Exploring Chemical Space and Drug Discovery*; Goodnow, R. A., Jr., Ed John Wiley and Sons: New York, 2014.

258. Franzini, R. M.; Samain, F.; Abd Elrahman, M.; Mikutis, G.; Nauer, A.; Zimmermann, M.; Scheuermann, J.; Hall, J.; Neri, D. Systematic Evaluation and Optimization of Modification Reactions of Oligonucleotides with Amines and Carboxylic Acids for the Synthesis of DNA-Encoded Chemical Libraries. *Biocon. Chem.* **2014**, *25*, 1453–1461.

259. Deng, H.; Zhou, J.; Sundersingh, F. S.; Summerfield, J.; Somers, D.; Messer, J. A.; Satz, A. L.; Ancellin, N.; Arico-Muendel, C. C.; Sargent Bedard, K. L.; Beljean, A.; Belyanskaya, S. L.; Bingham, R.; Smith, S. E.; Boursier, E.; Carter, P.; Centrella, P. A.; Clark, M. A.; Chung, C. W.; Davie, C. P.; Delorey, J. L.; Ding, Y.; Franklin, G. J.; Grady, L. C.; Herry, K.; Hobbs, C.; Kollmann, C. S.; Morgan, B. A.; Pothier Kaushansky, L. J.; Zhou, Q. Discovery, SAR, and X-ray Binding Mode Study of BCATm Inhibitors from a Novel DNA-Encoded Library. *ACS Med. Chem. Lett.* **2015**, *6*, 919–-924.

260. Franzini, R. M.; Biendl, S.; Mikutis, G.; Samain, F.; Scheuermann, J.; Neri, D. "Cap-and-catch" Purification for Enhancing the Quality of Libraries of DNA Conjugates. *ACS Comb. Sci.* **2015**, *17*, 393–-398.

261. Li, Y.; Zhao, P.; Zhang, M.; Zhao, X.; Li, X. Multistep DNA-Templated Synthesis Using a Universal Template. *J. Am. Chem. Soc.* **2012**, *135*, 17727–17730.

262. Clark, M. A.; Acharya, R. A.; Arico-Muendel, C. C.; Belyanskaya, S. L.; Benjamin, D. R.; Carlson, N. R.; Centrella, P. A.; Chiu, C. H.; Creaser, S. P.; Cuozzo, J. W.; Davie, C. P.; Ding, Y.; Franklin, G. J.; Franzen, K. D.; Gefter, M. L.; Hale, S. P.; Hansen, N. J.; Israel, D. I.; Jiang, J.; Kavarana, M. J.; Kelley, M. S.; Kollmann, C. S.; Li, F.; Lind, K.; Mataruse, S.; Medeiros, P. F.; Messer, J. A.; Myers, P.; O'Keefe, H.; Oliff, M. C.; Rise, C. E.; Satz, A. L.; Skinner, S. R.; Svendsen, J. L.; Tang, L.; van Vloten, K.; Wagner, R. W.; Yao, G.; Zhao, B.; Morgan, B. A. Design, Synthesis and Selection of DNA-Encoded Small-Molecule Libraries. *Nat. Chem. Biol.* **2009**, *5*, 647–654.

263. Satz, A. L. DNA Encoded Library Selections and Insights Provided by computational Simulations. *ACS Chem. Biol.* **2015**, *10*, 2237–2245.

264. Satz, A. L.; Hochstrasser, R.; Petersen, A. C. Analysis of Current DNA Encoded Library Screening Data Indicates Higher False Negative Rates for Numerically Larger Libraries. *ACS Comb. Sci.* **2017**, *19*, 234–238.

265. Kalliokoski, T. Price-Focused Analysis of Commercially Available Building Blocks for Combinatorial Library Synthesis. *ACS Comb. Sci.* **2015**, *17*, 600–607.
266. Leimbacher, M.; Zhang, Y.; Mannocci, L.; Stravs, M.; Geppert, T.; Scheuermann, J.; Schneider, G.; Neri, D. Discovery of Small Molecule Interleukin-2 Inhibitors From a DNA-Encoded Chemical Library. *Chem. - Eur. J.* **2012**, *18*, 7729–7737.
267. Melkko, S.; Mannocci, L.; Dumelin, C. E.; Villa, A.; Sommavilla, R.; Zhang, Y.; Grutter, M. G.; Keller, N.; Jermutus, L.; Jackson, R. H.; Scheuermann, J.; Neri, D. Isolation of a Small-Molecule Inhibitor of the Antiapoptotic Protein Bcl-xL From a DNA-Encoded Chemical Library. *ChemMedChem* **2010**, *5*, 584–590.
268. Terrett, N. In: *Moving in New Circles – An Introduction to Macrocycles in Drug Discovery*; *Drug Discovery Conference, CHI, San Diego, April*; 2014.
269. Maianti, J. P.; McFedries, A.; Foda, Z. H.; Kleiner, R. E.; Du, X. Q.; Leissring, M. A.; Tang, W. J.; Charron, M. J.; Seeliger, M. A.; Saghatelian, A.; Liu, D. R. Anti-Diabetic Activity of Insulin-Degrading Enzyme Inhibitors Mediated by Multiple Hormones. *Nature* **2014**, *511*, 94–98.
270. Du, C.; Fang, M.; Li, Y.; Li, L.; Wang, X. Smac, a Mitochondrial Protein That Promotes Cytochrome c–Dependent Caspase Activation by Eliminating IAP Inhibition. *Cell* **2000**, *102*, 33–42.
271. Varfolomeev, E.; Vucic, D. Inhibitor of Apoptosis Proteins: Fascinating Biology Leads to Attractive Tumor Therapeutic Targets. *Future Oncol.* **2011**, *7*, 633–648.
272. Riedl, S. J.; Renatus, M.; Schwarzenbacher, R.; Zhou, Q.; Sun, C.; Fesik, S. W.; Liddington, R. C.; Salvesen, G. S. Structural Basis for the Inhibition of Caspase-3 by XIAP. *Cell* **2001**, *104*, 791–800.
273. Mannhold, R.; Fulda, S.; Carosati, E. IAP Antagonists: Promising Candidates for Cancer Therapy. *Drug Discov. Today* **2010**, *15*, 210–219.
274. Sun, H.; Nikolovska-Coleska, Z.; Yang, C. Y.; Qian, D.; Lu, J.; Qiu, S.; Bai, L.; Peng, Y.; Cai, Q.; Wang, S. Design of Small-Molecule Peptidic and Nonpeptidic Smac Mimetics. *Acc. Chem. Res.* **2008**, *41*, 1264–1277.
275. Liu, Z.; Sun, C.; Olejniczak, E. T.; Meadows, R. P.; Betz, S. F.; Oost, T.; Herrmann, J.; Wu, J. C.; Fesik, S. W. Structural Basis for Binding of Smac/DIABLO to the XIAP BIR3 Domain. *Nature* **2000**, *408*, 1004–1008.
276. Gartner, Z. J.; Kanan, M. W.; Liu, D. R. Multistep Small-Molecule Synthesis Programmed by DNA Templates. *J. Am. Chem. Soc.* **2002**, *124*, 10304–10306.
277. Hennessy, E. J.; Adam, A.; Aquila, B. M.; Castriotta, L. M.; Cook, D.; Hattersley, M.; Hird, A. W.; Huntington, C.; Kamhi, V. M.; Laing, N. M.; Li, D.; MacIntyre, T.; Omer, C. A.; Oza, V.; Patterson, T.; Repik, G.; Rooney, M. T.; Saeh, J. C.; Sha, L.; Vasbinder, M. M.; Wang, H.; Whitston, D. Discovery of a Novel Class of Dimeric Smac Mimetics as potent IAP Antagonists Resulting in a Clinical Candidate for the Treatment of Cancer (AZD5582). *J. Med. Chem.* **2013**, *56*, 9897–9919.
278. Zhang, Y.; Seigal, B. A.; Terrett, N. K.; Talbott, R. L.; Fargnoli, J.; Naglich, J. G.; Chaudhry, C.; Posy, S. L.; Vuppugalla, R.; Cornelius, G.; Lei, M.; Wang, C.; Zhang, Y.; Schmidt, R. J.; Wei, D. D.; Miller, M. M.; Allen, M. P.; Li, L.; Carter, P. H.; Vite, G. D.; Borzilleri, R. M. Dimeric Macrocyclic Antagonists of Inhibitor of Apoptosis Proteins for the Treatment of Cancer. *ACS Med. Chem. Lett.* **2015**, *6*, 770–775.

FURTHER READING

279. Friedman, R. Aggregation of Amyloids in a Cellular Context: Modeling and Experiment. *Biochem. J.* **2011**, *438*, 415–426.

Index

Note: Page numbers followed by *f* indicate figures.

A

Activity-based protein profiling (ABPP),
  50–52, 51–53*f*
Affinity selection (AS-MS), 100–101
Amphotericin B, 138–140
Aqueous solubility, 123
ATP-binding domain, Hsp90, 125

B

BET. *See* Bromodomain and extraterminal
  domain (BET)
Biology-oriented synthesis (BIOS), 136, 140*f*
Bioluminescence resonance energy transfer
  (BRET), 81–82
Bromodomain and extraterminal domain (BET)
  affinity chromatography, 61–62
  benzodiazepine, 60
  biological profiling, 63–64
  characterization, 64
  I-BET, 61
  identification, 60, 61*f*
  oncology, 64
  proinflammatory gene expression, 60
  selective BET reader inhibitor, 61–62*f*
  X-ray complexes of Brd proteins, 62, 63*f*
Bruton's tyrosine kinase (BTK), 50–52

C

Capture compound mass spectrometry
  (CCMS), 52
Cellular quality control system, 125
Chameleon effect, 144, 144*f*
Chaperones, 125
Chemical diversity (CD) collections, 79
Chromatin
  BET, 60–64, 61*f*, 63*f*
  heterochromatin, 57
  histone proteins, 56
  histone PTMs, 58–59, 59–60*f*
  packing, 57–58, 58*f*
  structure, 56, 57*f*

Combinatorial chemistry/combichem, 117,
  119, 136
Computer-assisted drug design (CADD), 84
CRISPR-Cas9 gene editing
  process and main components, 8, 8*f*
  TV models, 35
Cyclosporine, 138–140

D

DECLs. *See* DNA-encoded libraries (DECLs)
Differential scanning fluorimetry (DSF), 100
Direct fluorescence measurements, 81–82
Discrete libraries, 119
Diversity-oriented synthesis (DOS), 136, 139*f*
DNA-encoded chemistry and affinity
  screening, 153
DNA-encoded libraries (DECLs)
  affinity-based screening, 153–155
  chemical transformations, 155, 156*f*
  diversity and size, 156–157
  drug-likeness, 157
  5-BB macrocyclic, 158
  libraries quality, 156
  linear reaction sequences, 156
  mix-and-split synthesis, 153, 154*f*
  monofunctional reagents, 156
  P_2–P_5 library, 158, 160
  scaffolds, BBs, and build-up strategies, 156,
    157*f*
  screening strategy, 153, 155*f*
  SS DNA-BBs' libraries, 153
  SS DNA complementarity, 153
  structures, 157, 158*f*
  targeted against IAPs, 158, 159*f*
  2-BB DECLs, 157
DNA-templated chemistry (DTC), 153
Drug affinity response target stability
  (DARTS), 40, 41*f*
Drug-like chemical diversity
  active principles isolation, 115–116
  compound collections
    access, 117

Drug-like chemical diversity *(Continued)*
    assay interference, 121
    assembled collections, 124
    assembly, 117
    extended RO5, 120, 120*f*
    FBDD collections, 123–124, 123*f*
    HTS collections, 117, 118*f*, 119
    iterative organic synthesis, 117
    lipophilicity, 119–120
    medium-large compound collections, 117
    mix-and-split synthesis, 117–119
    molecular fragments, 122
    MW, 119–120, 122
    numerosity *vs.* diversity, 117–118, 118*f*
    organic scaffolds, 118–119
    PAINS, 121–124, 121*f*
    parallel synthesis, 119
    quality, 117
    RO5 *(see* Rule of 5 (RO5))
    RO3 filter, 122, 122*f*
    targeted libraries, 119
    Veber, 120, 120*f*
    vHTS, 119
  drug discovery paradigm, 115, 116*f*
  Hsp90 *(see* Heat shock protein 90 (Hsp90))
  natural remedies, 115, 117
  purification techniques, 115–116
  R&D pharmaceutical process, 115, 116*f*

E

Electrochemiluminescence, 81–82
Empirical scoring functions, 86
Encoded self-assembling libraries (ESACs), 153
Ensemble docking, 87
Epothilone B, 138–140
Erythromycin, 138–140
Extended RO5 (eRO5), 120, 120*f*

F

False negatives, 118–119
False positives, 118–119
FBDD. *See* Fragment-based drug discovery
    (FBDD)
FCG. *See* Forward chemical genetics (FCG)
[19]F-labeled reporter screen ligand NMR, 101–102
Flow cytometry, 81–82
Fluorescence detection, 81–82, 82*f*
Force-field-based scoring functions, 86
Forward chemical genetics (FCG), 10–12, 12*f*
    ABPP, 50–52, 51–53*f*
    affinity chromatography, 41–42, 42*f*

bio-orthogonal probes, 49, 50*f*, 53, 54*f*
biotin, 52
chemical labeling, 47, 48*f*
chemical proteomics, 41
DARTS, 40, 41*f*
endogenous molecules, 36
expression-cloning techniques, 54–56, 55*f*
functionalization, 42
gene dosage
    HIP, 38, 39*f*
    HOP, 38, 39*f*
    MSP, 38, 39*f*
iTRAQ, 49
JAK-STAT signaling, kinase-independent
    inhibition, 42, 43*f*
metabolic labeling, 47, 47*f*
prefractionation, 45
proteomic profiling, 40
putative targets, 53–54
RNAi, 40
soluble matrix-free affinity reagents, 49
specific *vs.* aspecific immobilized affinity
    reagent-protein interactions, 45, 46*f*
SPROX, 40–41, 41*f*
tangible target deconvolution methods, 36,
    36*f*, 38
TICC, 41
transcriptomic/mRNA profiling, 40
trifluoromethyl diazirine linkers, 44–45, 44*f*
trifunctional probes, 49
virtual target deconvolution methods, 36–38,
    36–37*f*
Y3H, 56
Fragment-based drug discovery (FBDD)
    biochemical HTS assay formats, 98–99,
        99–100*f*
    collections, 122–124, 123*f*
Functional cloning, 3
Function-oriented synthesis (FOS), 136

G

Gain-of-function (GOF) gene knockin, 35

H

Haploin-sufficiency profiling (HIP), 38, 39*f*
HD. *See* Hit discovery (HD)
Heat shock protein 90 (Hsp90)
    ACD database, 129–131
    aminopyrimidine, 132
    chaperones, 125
    compound VER-49009, 134–135

cytoplasmic Hsp90 isoforms, 125
domains, 125
hit progression at Vernalis, 132, 133*f*
HSF1 release, 126
isoxazoles, 135
kinase-directed SeeDs 3 sublibrary, 131
lead progression at Vernalis, 133, 134*f*
medchem-based filters, 128, 128*f*
misfolded dysfunctional proteins, 124
MW filter, 129
natural and synthetic Hsp90 inhibitors, 126,
    127*f*
NBD, crystal structures, 128
NMR fragment screening, 129, 130–132*f*,
    131–133
in oncology, 126
PQC, 125
and PU3
    analogues, 126–128
    binding, 128
    molecular interactions, 126, 127*f*
    structural optimization, 128
resorcinol pyrazoles, 133–135
STD, 131
thienopyridine, 132
vendors and filters, 128, 129*f*
virtual screening, 128–129, 130*f*
water-ligand observed gradient spectroscopy,
    131
HedgeHog-Interacting Protein (HHIP), 150
Heterochromatin, 57
High-throughput screening (HTS), 34, 77, 79,
    115
biology-oriented HTS
    academic HTS, 82
    assay formats, 81–82, 82*f*
    cell-free/target-based assays, 80–81, 81*f*
    cellular/phenotypic assays, 80–81, 81*f*
    costs, 83
    in vitro screening, 79
    microfluidics, 80
    miniaturization, 79–80
    modular automation, 79–80
    penicillin discovery, 79
    private and public HTS, 82–83
    robustness and compliance, 79–80
    standardization, 79
    whole organism assays, 80–81, 81*f*
biophysical screening
    biophysical assay formats, 99, 100*f*
    drug-like/bioavailable small molecules, 98
    FBDD, 98–99, 99–100*f*
    fragment collections, 100, 102–104
    ligand-target and fragment-target fits,
        98, 99*f*
    methods, 100
    MS methods, 100–101
    nanodrop crystallization, 102
    NMR methods, 101–102
    orthogonal sequential screening, 102, 104*f*
    primary screening, 102–104
    SPR, 101
    TSA-based formats, 100
    X-ray crystallography, 102
    vHTS (*see* Virtual HTS (vHTS))
Histone marks, 58–59, 59–60*f*
Histone PTMs, 58–59, 59–60*f*
Hit discovery (HD), 115
    CD collections, 79
    drug discovery, 79
    high-quality hits, 77
    HTS (*see* High-throughput screening (HTS))
    interdisciplinary nature, 77
    physicochemical properties, 78–79
    R&D pharmaceutical process, 77, 78*f*
    valuable hits, 77–78, 78*f*
Homozygous deletion profiling (HOP), 38, 39*f*
Hsp90. *See* Heat shock protein 90 (Hsp90)
HTS. *See* High-throughput screening (HTS)
Human epidemiological data, 35
Hydrogen bond acceptor (HBA), 119–120
Hydrogen bond donor (HBD), 119–120
Hydrogen-deuterium exchange (HDX-MS),
    100–101

I

ICAT. *See* Isotope-coded affinity tag (ICAT)
Inhibitor of Apoptosis proteins (IAPs),
    157–158, 159–161*f*, 160
Investigational New Drug (IND) application, 9
Isobaric relative and absolute tag for
    quantification (iTRAQ), 49
Isothermal calorimetry (ITC), 100
Isotope-coded affinity tag (ICAT), 47–49, 48*f*
Iterative organic synthesis, 117

K

Kinesin spindle protein (KSP)
    allosteric binding site, 22, 22–23*f*
    anaphase, 14–15
    cell-based cytoblot assay, 17
    cell-cycle-regulating kinases, 15
    cell growth and protein synthesis, 12–13

Kinesin spindle protein (KSP) *(Continued)*
  cytokinesis, 15
  druggable target, 22
  human Eg5-directed mAb, 20
  interphase process, 12–13
  karyogenesis, 13
  kinase inhibitors structure, 15, 16*f*
  kinesin Eg5 and MTs interactions, 21
  mammalian cell cycle, 12, 13*f*
  mammalian mitosis, 13, 14*f*
  metaphase, 14
  mitotic spindle, 19, 20*f*
  molecular motors, 15
  monastrol, 19–22
  MTAs structure (*see* Microtubule-targeting
    agents (MTAs) structure)
  phenotypic screening-centered strategy,
    15–17, 17*f*
  PLK inhibitors, 15
  primary HTS assay, 17
  prometaphase, 13–14
  racemic mixture, 19
  telophase, 14–15
  Xenopus egg, 19–20
Knowledge-based scoring functions, 86
KSP. *See* Kinesin spindle protein (KSP)

L

Label-free technologies, 81–82
Ligand-based drug design (LBDD), 89
Ligand-bound SPR, 101
Ligand-observed NMR, 101
Lipophilicity, 119–120
Loss-of-function (LOF) gene knockout, 35
Luminescence, 81–82

M

Macrocycles
  amide, 155
  amphotericin B, 138–140
  aqueous environments, 144–145
  bioactive macrocyclic NPs, 138, 141*f*
  bioavailability, 144, 144*f*
  biosynthetic macrocycles, 145, 149, 151,
    151*f*
  CP2 cocrystallization with KDM4A, 153
  cyclosporine, 138–140
  DECLs (*see* DNA-encoded libraries
    (DECLs))
  dimeric Smac mimetics, 158, 160
  DTC, 153

  epothilone B, 138–140
  erythromycin, 138–140
  ESACs, 153
  IAPs, 157–158, 159–161*f*, 160
  lead, 162
  libraries, 148, 149*f*, 151, 151*f*
  linear small molecule, 143
  macrocyclizations, 145–146
  macrolactamization, 145–146
  modified translational codon code, 151–152
  monomers, 158
  natural, 138, 140–142, 141–142*f*
  O2beY, 149, 151
  PPI inhibitors, 142–143
  P_2-P_5 macrocyclic library, 158, 160
  RaPID, 152, 152*f*
  rifamycin SV, 138–140
  Shh recognition sequence, 150
  SS DNA complementarity, 153
  synthetic, 138, 145–148, 145–148*f*
  tangible collections, 148
  virtual collections, 147–148
  XIAP BIR2 complex, 160
Mammalian cell cycle, 12, 13*f*
Mechanism of action (MoA), 2–3
MicroRNA (miRNA), 6*f*, 7
Microtubule-targeting agents (MTAs) structure
  colchicine, vinblastine, taxol, 15, 16*f*
  with lower potency, 18
  mitosis-targeted phenotypic screening
    group VII compounds, 19, 19*f*
    nocodazole, tetrahydrothiazepines,
      synstab, 17–18, 18*f*
  non-MTA positives, 18
Mix-and-split synthesis, 117–119
Molecular fragments, 122
Molecular weight (MW), 119–120, 122
MTAs structure. *See* Microtubule-targeting
  agents (MTAs) structure
Multicopy suppression profiling (MSP),
  38, 39*f*

N

Nanodrop crystallization, 102
Native electron spray ionization (Native
  ESI-MS), 100–101
Natural products (NPs), 9
  BIOS, 136, 140*f*
  combichem, 136
  DOS, 136, 139*f*
  libraries, 136, 138–139*f*
  macrocycles (*see* Macrocycles)

modulators, 135–136
penicillin, 135–136
scaffolds, 136
selected and semisynthetic derivatives, 136,
    137*f*
target validation, 34
Nearest-neighbor search (*n*-(NN)), 94, 94*f*
Neurodegenerative diseases (NDs), 9–10
NPs. *See* Natural products (NPs)
Nucleosomes, 56–57
Numerosity, 117–118, 118*f*

O

O-2-bromoethyl tyrosine (O2beY), 149, 151
Organic scaffolds, 118–119
Orthogonal sequential screening, 102, 104*f*

P

Pan assay interference compounds (PAINS),
    121–124, 121*f*
Photoaffinity labeling, 52
Photoaffinity probes (PAPs), 52, 52*f*
Polo-like kinase (PLK) inhibitors, 15
Polymerase chain reaction (PCR), 3
Positional cloning, 3, 3*f*
Pre-Watson and Crick drugs, 2–3
Protein-bound SPR, 101
Protein complementation assays (PCA), 81–82
Protein-observed NMR, 101, 102
Protein-protein interactions (PPIs), 112
Protein quality control (PQC), 125
Proteomic profiling, 40

R

Radioactivity detection, 81–82, 82*f*
Resorcinol pyrazoles, 135
Rhodamines, 121–122
Rifamycin SV, 138–140
RISC processing, 7
RNA interference (RNAi), 5–7, 6*f*, 35
Rotamer library sampling, 87
Rule of 3 (RO3), 122, 122*f*
Rule of 5 (RO5)
    definition, 119, 120*f*
    Lipinski's RO5, 122
    NPs
        BIOS, 136, 140*f*
        combichem, 136
        DOS, 136, 139*f*
        libraries, 136, 138–139*f*

macrocycles (*see* Macrocycles)
        modulators, 135–136
        penicillin, 135–136
        scaffolds, 136
        selected and semisynthetic derivatives,
            136, 137*f*
    RO5-Veber filters, 120
    soft RO5 versions, 120
    WDI, 119

S

Saturation transfer difference (STD), 131
Scintillation proximity assays (SPA), 81–82
Side chain flexibility, 87
SILAC. *See* Stable isotope labeling by
    aminoacids in cell culture (SILAC)
Single-stranded (SS) DNA-BBs' libraries, 153
Small interfering RNA (siRNA), 6*f*, 7
Sonic Hedgehog (Shh) recognition sequence, 150
Stability of proteins from rates of oxidation
    (SPROX), 40–41, 41*f*
Stable isotope labeling by aminoacids in cell
    culture (SILAC), 47, 47*f*
Structure-activity relationships (SARs), 80–81
Structure-based drug design (SBDD), 87, 89
Structure-based pharmacophores, 97
Surface plasmon resonance (SPR), 101

T

Tanimoto coefficient, 92
Tanimoto index, 37
Targeted libraries, 119
Target identification (TI)
    chemistry
        anti-inflammatory agent, 9
        biology-driven TI, 10, 11*f*
        data-mining-mutated genes, 10
        drug-like chemical diversity, 11, 17*f*
        genome-sequencing-mutated genes, 10
        HT genome sequencing and identification,
            10
        IND application, 9
        mutated-gene-containing cell lines, 10
        mutated-gene-containing zebrafish, 10
        NDs, 9–10
        NPs, 9
        -omics revolution, 9
        phenotype-based approaches, 9
        phenotypic screening, 10
        phenotypic screening-based FCG,
            10–12, 12*f*

Target identification (TI) *(Continued)*
   prostaglandin synthesis, 9
   RNAi and CRISPR-induced gene editing,
     10, 11*f*
   small-molecule modulator-molecular
     target connection, 11
   target-based approach, 9
 disease-related target, 1
 druggable genome, 1, 2*f*
 KSP (*see* Kinesin spindle protein (KSP))
 molecular biology
   functional cloning, 3
   gene-encoded protein targets, 2–3
   genome-wide gene editing, 8, 8*f*
   genome-wide RNAi libraries, 7–8
   massive parallel DNA sequencing, 3*f*, 4
   miRNA, 6*f*, 7
   MoA, 2–3
   -omics revolution, 4
   polymerase chain reaction, 3
   positional cloning, 3, 3*f*
   pre-Watson and Crick drugs, 2–3
   recombinant DNA, 3
   RNAi, 5–7, 6*f*
   siRNA, 6*f*, 7
   systems biology, 4–5, 4–5*f*
   target-related efficacy and safety, 3
 putative drug target genes, 1
 R&D pharmaceutical process, 1, 2*f*
Target identification by chromatographic
   coelution (TICC), 41
Target validation (TV)
 BET (*see* Bromodomain and extraterminal
   domain (BET))
 chemistry (*see* Forward chemical genetics
   (FCG))
 chromatin (*see* Chromatin)
 definition, 33
 innovative mechanisms, 33
 molecular biology
   biology-driven validation methods, 35
   cloning, 34–35
   irreversible TV models, 35
   massive parallel sequencing, 34–35
   natural products and extracts, 34
   -omics revolution, 34–35
   physiological and traditional-medicine-
     driven observations, 34
   systems biology, 34–35
   target-disease connection, 34
 Phase II/Phase III clinical failure, 33–34
 R&D pharmaceutical process, 33, 34*f*

Thermal shift assay (TSA), 100
3D-molecular fingerprints, 90–92, 94–96*f*, 95–97
3D-pharmacophores, 90–92, 94–96*f*, 95–97
Transcriptomic/mRNA profiling, 40

V

Virtual HTS (vHTS), 119
 CADD, 84
 computer-based methods, 83
 de novo modeling, 84–85
 dynamic targets and small-molecule ligands,
   86–87
 false positives, 89
 fingerprint recombination, 94*f*, 95
 fold recognition, 84–85
 homology modeling, 84–85
 information-driven virtual screening, 87, 87*f*
 in silico protein flexibility, 87
 in silico target model, 84
 LBDD, 89
 molecular descriptors, 85–86, 89
 molecular docking, 86, 86*f*
 nearest-neighbor search, 94, 94*f*
 1D-, 2D-, and 3D-features, 85, 85*f*
 SBDD, 87, 89
 scoring functions, 86
 structure-based pharmacophore, 89
 structure-based vHTS, 88, 88*f*
 tangible positives, 88–89
 Tanimoto coefficient, 92
 3D-pharmacophores, 90–92, 94–96*f*, 95–97
 2D-molecular fingerprints, 89
   circular paths, 90, 92*f*
   hybrid fingerprints, 90, 93*f*
   substructure-based 2D fingerprints, 89, 90*f*
   topology-based fingerprints, 89–90, 91*f*
 virtual libraries, 88
 virtual target structure, 84, 84*f*
Virtual pharmacological space, 37, 37*f*

W

Wild-type (WT) cell genes, 38
World Drug Index (WDI), 119–120

X

X-ray crystallography, 102

Y

Yeast three-hybrid systems (Y3H), 56

Printed and bound by CPI Group (UK) Ltd, Croydon, CR0 4YY

03/10/2024

01040399-0019